History of the Eastern Railways Construction and Expansion

History of the Eastern Railways Construction and Expansion

The Forgotten Journey

VOLUME I

DOMINIC WICKS

Copyright © 2022 by Dominic Wicks

All rights reserved. No part of this book may be reproduced or used in any manner without written permission of the copyright owner except for the use of quotations in a book review. For more information, address: dk.wicks41@gmail.com

FIRST EDITION

ISBN 978-1-80227-408-0 (paperback)
ISBN 978-1-80227-409-7 (ebook)

Contents

Preface	vii
1 The Preparatory Work and the Politics of Construction	1
2 The Construction of the Eastern Railway	5
3 Stargard-Posener Railway Company	8
4 Eastern Railway Lines West and East of the Vistula	10
5 Eastern Railway Connections at Nogat and Vistula Bridges	12
6 Copy of an Early Railway Timetable for 1855	14
7 Completion of Eastern Railway Connections from West to East Prussia	15
8 Further Development on the Eastern Railway	17
9 Eastern Railway Network 1875-1876	20
10 Frontier Stations and the Border Routes of the Eastern Railway	22
11 Earliest Known Copy of Prussian Railway Tariffs	24
12 International Agreement between Prussia and Russia	25
13 The Eastern Railway's Continuous Connections to the Russian Border 1862	29
14 The Eastern Railway's Connections at the Polish Border	30
15 'Fährpost'	32
16 The Prussian Travelling Post Office	34
17 Use of the Mailbox on the Railway Mail Cars	40
18 Emergence of Freight Mail Cars on the Eastern Railway	43
19 Eastern Railway Main Lines between Berlin and East Prussia 1851-1889	48

20	Eastern Railway Route Postmarks	49
21	The Prussian Railway Post Offices in Berlin	54
22	Berlin Main Stations – Earliest Concentric Postmarks	56
23	Imperial Railway Post Office Berlin Receiving Foreign Mail	62
24	Berlin Stettiner Railway Station for the Postspeditonsamt III (Ostbahn Routes)	63
25	Stettin Railway Station Office Postmark	68
26	Line Stamp Railway Postmarks of Stettin and Postmark Catalogue	70
27	Railway Office: Station (City) Expedition	73
28	Railway Station Postmark – SPED.COMTOIR No.3 Stettin	74
29	Eisenbahn Post Büreau 3 Stettin (Deutsche Reich)	78
30	Small Oval Postmarks of Stettin 1883-1921	80
31	Family Escape in the Rail Mail Car – 1945	88
32	Withdrawal, Surrender and Dissolution of Reich Railway Management Stettin	91
33	Railway Station Postmarks – Deutsche Reich Period 1889-1945	97
34	Railway Mail at Frankfurt Oder	98
35	Railway Construction in and around Kreuze Station	109
36	Railway Postmarks of Cüstrin Station	112
37	Prussian Eastern Railway Station Schneidemühl	115
38	Berlin – Cöslin – Stolp – Danzig	122

THE STEAM LOCOMOTIVES

39	Historical Development of Prussian-Era Steam Locomotives	126
40	The Prussian D-ZUG Locomotives of the Eastern Railway	130
41	The 2-B-1 Express Locomotive – Prussian State Railway P 4	132
42	Prussian State Railways Locomotive Class S 2	135
43	Prussian State Railways – Locomotive Series T 8	138
44	The Deutsche Reich Railway (DRG) Series 43	141
45	The Deutsche Reich Railway Company Class 24 (Sketch)	143

Preface

A *History of the Eastern Railway* (the "Ostbahn") rediscovers the lost railway lines of Prussia and their locomotives. It is also about my interest in collecting little-known railway mail postmarks from the period 1848 to 1945.

The first chapter of this volume explores the politics and the construction of a state-owned railway, as well as the designs and drawings of the first parts of the route by early pioneers of railway engineering.

Later chapters trace a railway mail journey by steam train from Berlin to the Russian border and beyond, into the occupied Baltic States and Ukraine.

The route network supported the longest uninterrupted railway route, with a series of key and minor stations, including branch lines connecting the lake routes of West and East Prussia.

We explore some of the activities at the key stations, the birth of the railway companies, their management, and the amalgamations and nationalisation of private routes, from 1852 to 1889. From 1889 to 1945 we see the changes to railway management as a result of wars, culminating in the companies' surrender and dissolution.

I have provided detailed descriptions of some of the steam locomotives that ran on the network, including the narrow-gauge steam locomotives.

The written content and the material presented here are appearing for the first time in the public domain. Contributors include knowledgeable members of the Railway Post Office Society in Germany.

The material images presented are either from my collection or from the Archives of the Bundesarbeitsgemeinschaft Bahnpost e.V. Frankfurt Study Group.

References to the railway mail postmark (BArGe) catalogue, as well as Michel Deutschland catalogue numbers, are used.

The Preparatory Work and the Politics of Construction

1

The Prussian Minister August von der Heydt, who became Minister of Commerce at the end of 1848, passed a law on 7 December 1849 by which the following railway lines would be state-owned: the Eastern Railway, the Westphalian Railway and the Saarbrücken Railway. These were the earliest state railways.

As early as 5 November 1848, the Royal Management of the Eastern Railway was formed and took charge of planning and the construction process. The negotiations were led by boards of directors (organised into four classes: princes and noblemen; knights of the Principality; municipalities; rural communities). They advised on various routes, fieldwork, bridge construction and fundraising. Among other things, the question of finance was discussed and where the crossing of the Vistula should be, as that was a critical aspect of the line's design. A bridge was initially planned at Graudenz, but the decision was finally made to cross over at Dirschau and Marienburg, although here not only the Vistula but also the Nogat River had to be bridged.

August von der Heydt (1801–1874)

The main highways from Berlin to Konigsberg passed through West Prussia and continued in the direction of Dirschau. A decision to build a branch line to Danzig was achievable in Dirschau as it was the shortest connection. The necessary regulations for the crossing of the Vistula and Nogat not only benefited the railway, but the whole country. The Nogat and Marienwerder lowlands were protected by spillway dams in the river from flooding, at the same time construction of a number of storage reservoirs and spillway dams in the river and its tributaries increased water inflow the Vistula and improved navigation due to the railways and the bridge construction that followed.

2 HISTORY OF THE EASTERN RAILWAYS CONSTRUCTION AND EXPANSION

Railway engineers then planned their next move, crossing the Elbing River. Their canals connected Braunsberg with a final connection to Königsberg. Preliminary negotiations highlighted military reasons. National defence was an essential motive for the construction of the railway. It was then pointed out that just at a time when "the defence of civilization and the blossoming constitutional freedom of Prussia and Germany were to take place, it was not possible to build a 'path to peace' but to consider the possibility of war." In contrast to the eastern part of the railway line from Dirschau to Königsberg, beating a path were branch connections to economically important towns. Amongst the planned connections, the Cüstrin station formed a junction with the private Breslau-Stettin railway line. There was a direct connection from Berlin via Cüstrin to stations further afield at Landsberg (Warthe), Driesen in Ostbrandenburg, Bromberg and Dirschau.

A short distance from Driesen, the proposed route was crossed by the Stargard-Posen private railway. Due to insufficient funds to complete the expansion of the entire route from Berlin to Bromberg, it was decided the route from Berlin to Driesen would be left until such time as funds were available. The route was to terminate at the intersection with the existing Stargard-Posener Railway near Driesen.

As the funds for the construction were made available following political agreements, the eastern route continued to expand towards Schneidemühl and on to Bromberg. As a result, Berlin was finally connected via the Stettin-Posen Railway with the Eastern Province.

The land acquisition for the Eastern Railway line was the responsibility of the railway management. The engineering teams planned a double track to cope with population density, but due to limited funds, it was initially built as a single railway line. The negotiations for the extension of the route took a long time because of disagreements in the National Assembly on the method of raising funds. Each section of the route had to be approved by the National Assembly as it crossed international borders.

It was not until the political turmoil of 1848 (the March Revolution in Germany and Austria aimed at forming a democratic constitution) and the resulting unfavourable working conditions in Germany, especially in Berlin, that the Prussian minister, von Heydt, was prompted to begin the construction of the Eastern Railway.

A new branch line was proposed which would shorten the distance and time it took early steam trains to reach Bromberg. This new line would cross the existing Posner-Stargarder Railway (opened in 1847/48). The proposed line would also form a branch for trains from Posen via Kreuz and a separate branch connection to Schneidemühl. For the railway company, these were the first branch routes.

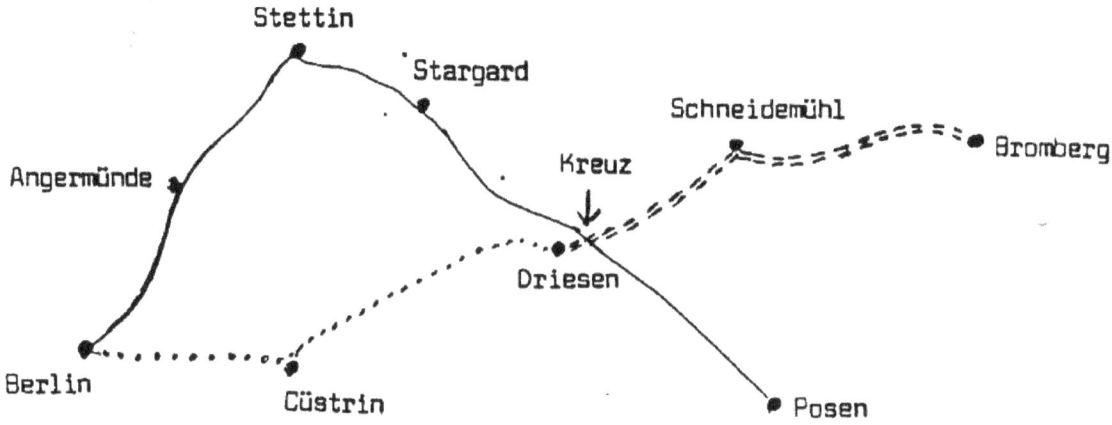

Original Proposal of the route construction without a station stop at Kreuz

The proposed route from Berlin via Cüstrin and Kreuz Station to Bromberg was completed on 27 July 1851 and proved popular with engineering companies supplying material for further extensions. The route sketch supports two railway lines crossing at Kreuz. The station was the crossing point for the line to Posen. The sketch shows Kreuz town with a single rail line at first and the station. A second route between Stettin and Posen was built by a private railway company.

During mid-nineteenth-century wars with Germany's neighbours, the railways played a pivotal role for the Prussian Army. The Prussian government placed heavy demands on private railway companies for their war effort, leading to fewer trains for public use. The development plans constantly changed due to the operating nature of the Prussian privately operated network. This was overshadowed by the forceful nature of changes imposed by none other than the larger-than-life figure of Paul von der Heydt, who used every lever of the state's legal and financial powers to give the state the upper hand in railway ownership. Heydt's aggressive tactics jeopardised one of Europe's largest investment sectors and antagonised Prussia's leading entrepreneurs. Another example of his authoritarianism was forcing private railway companies to run a night train service (which, in the end, proved a profitable avenue).

In the early years, Germany did not have a manufacturing facility for steam locomotives. Robert Stephenson and Company built a number of Crampton-type locomotives for the Prussian Eastern Railway. From 1851, a total of 157 locomotives were sold to Prussia. These all had a 4-2-0 wheel arrangement, with inside cylinders and indirect drive. The inside cylinders drove a crankshaft located in front of the firebox, and the crankshaft was coupled to the driving wheels by outside rods. In the long term, this solution seems less important than Crampton's adoption of wide steam passages, generous bearing surfaces and large heating surfaces, and it was these three features,

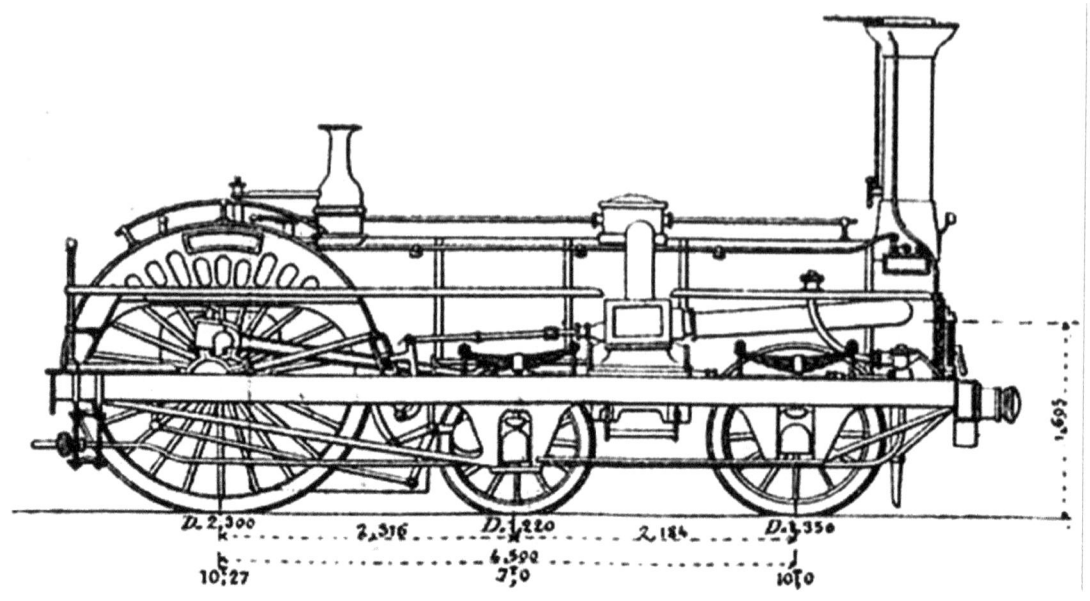

Drawing of Crampton steam locomotive built 1851–1852

rather than the position of the driving axle, that ensured the high-speed qualities of his locomotives.

Another peculiarity of some of Crampton locomotives was the use of a boiler of oval cross-section to lower the centre of gravity. It would nowadays be regarded as bad engineering practice because the internal pressure would tend to push the boiler into a circular cross-section and increase the risk of metal fatigue.

Crampton locomotives were used by some British railway lines, and speeds of up to 120 km/h (75 mph) were achieved on the London and North Western Railway (LNWR). They were more popular in France, southern Germany and the US. Over 300 were built for use on the Continent between 1846 and 1864. Some were running on the Prussian Eastern Railway until well into the 20th century.

The Construction of the Eastern Railway

2

For the construction of the Eastern Railway, the "Royal Management of the Eastern Railway" was used by the Prussian king Friedrich Wilhelm IV. It had all the powers of a public authority in matters relating to the business entrusted to it and was directly answerable to the Minister of Trade, Commerce and Public Works. This was the first Royal Railway Management in Prussia.

The Royal Management of the Eastern Railway took charge of the work from June 1848, with an office in Bromberg, for the construction of the line from Driesen to Bromberg, under the instructions of the "Royal Commission for the Eastern Railway". The management re-employed those who had been employed at Elbing in the construction of the Marienburg-Braunsberg route, District Administrator Wernich and Inspector Gerhardt from the buildings department.

On 26 June, the first Berlin workers' team of 204 men was transported from Dragebusch in Driesen in East Brandenburg to Kreuz. On 29 June, the existing team of railway builders was joined by another 100 men. For security, the military was located near the workplaces. The fears that these people would commit riots, as in Berlin, were not realised. Apart from local workers and a few Silesians, the largest group of settlers were from Berlin, and they were book printers, typesetters, bookbinders, painters, pattern makers, servicemen, fruit dealers, carpenters and others. The supervisory managers struggled with the heavy work of the cumbersome working carts, and as a result, there was a constant change of workers as the work progressed.

In September 1848 at Netzthal, Pomerania, the construction of the route began, with the earth bed streaked with hundreds of "finger lakes", streams flowing in and out on their way to one of the two rivers that had to be crossed to reach Nakel station. Discoveries were made on almost the whole stretch of the Eastern Railway. Amber was found on the bottom of a lake, and a lively trade flourished as a result. The centre of this trade was, for a time, Nakel, and from here contacts extended as far as Krakow, Bucharest and Constantinople.

While the Kreuz-Bromberg route was almost complete, the construction of the Bromberg-Danzig line was also underway. The number of workers increased steadily and amounted to 12,250 men in total by June 1851.

When the first section of the Eastern Railway was opened, it was recognised as the first long-distance state railway line in Prussia. King Friedrich Wilhelm IV wanted to take the journey himself! In addition, an electric telegraph line connecting the Eastern Railway to Königsberg and Danzig with the capital, including the cities of Stettin, Kreuz, Posen and Bromberg, was completed on 15 October 1849. Originally, the festival celebrations were planned for 31 July but were postponed until October when the completed stations were handed over to ordinary traffic. The stations were Filehne, Schönlanke, Schneidemühl, Miaseczko (Friedheim), Bialosliwe (Weißenhöhe), Osiel (Netztahl), Nakel and Bromberg.

Once completed, the Eastern Railway line opened up the route from Berlin to the Prussian provinces. The first public passenger train on this route left on 26 July 1851 at 11.00 p.m. from Stettiner station in Berlin. The first train from Bromberg in the east, which departed in the early hours of 27 July 1851 (2:30 a.m.) was empty.

Due to financial difficulties, the Stettin-Stargard-Posener railway line, which crossed the Eastern Railway line, was taken over by the Prussian administration, and its operations provided a comprehensive service with improved timetables to freight and passenger trains.

Thus, a continuous connection from Berlin to Bromberg was created, albeit initially with a considerable detour via Stettin.

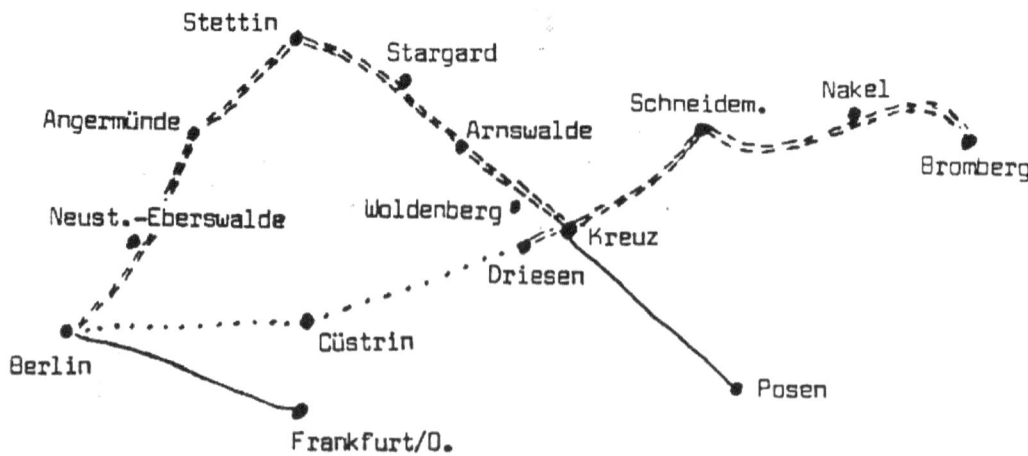

Proposed connections from Berlin to Bromberg

Operationally, the Stettin-Stargard-Kreuz-Bromberg route was considered a continuous line (see map above), joined by a branch line to Posen. The Eastern Railway, therefore, was able to offer continuous wagon trains between Stettin and Bromberg. The Stargard-Posener Train Company offered alternative services, with a change of trains at Kreuz for Posen for travellers from Berlin or Bromberg.

Stargard-Posener Railway Company

3

Founded in 1846, the Stargard-Posener Railway Company did not meet expectations in the beginning. The state intervened and transferred the management of the company to the Royal Management of Eastern Railway in Bromberg in 1851, then, in 1857, to the Upper Silesian Railway, which was also temporarily managed by the state. The final date of nationalisation was 1 January 1883.

STARGARD-POSENER RAILWAY TIMETABLE

The train company operated a daily service on the Ostbahn between Berlin and Bromberg. Two trains, which also took freight cars, ran in each direction at 35 km/h. The trains travelled according to the following timetable:

	DAY	NIGHT		DAY	NIGHT
Berlin	From 12.00	11.00	Bromberg	From 2.30	3.00
Stettin	From 3.55	2.50	Kreuz	From 6.36	7.24
	From 4.24	3.24		From 7.11	8.13
Kreuz	From 8.04	6.37	Stettin	From 10.47	11.45
	From 8.49	6.55		From 12.05	12.20
Bromberg	From 1.30	11.00	Berlin	From 4.04	4.32

On 15 October 1851, a delay of 15 minutes was set for the trains leaving from Bromberg. In the interwar period from 1918/19 to 1939, the line crossed the German-Polish border; the border stations were Kreuz and Drawski Młyn.

On this first Eastern Railway line from Bromberg to Kreuz and Stettin to Berlin, the black two-line postage stamp "OSTBAHN / Date – Tour/Trip number" was used, without numerals for the year, which was stamped on the back of the letters.

```
OSTBAHN
24 10 ~ III. R
```

For the Eastern Railway Bromberg-Berlin route, a reorganisation of the management was mentioned in the Official Journal, Issue 27, dated 6.10.1857. The following changes were to take place:

The railway post office (**Eisenbahn-Postamt**) No. IV was given the responsibility for the Bromberg-Königsberg section of the route.

The railway post office (**Eisenbahn-Postamt**) No. XI was moved from Danzig to Bromberg.

The current management of the Cöslin-Berlin route was the railway post office (**Eisenbahn-Postamt**) No. III.

The Breslau-Posen-Kreuz branch route was the responsibility of the railway post office (**Eisenbahn-Postamt**) No. XIV.

The rail line from Breslau via Posen and Kreuz to Stettin did not belong to the "classic" Eastern Railway line due to new connections.

Eastern Railway Lines West and East of the Vistula

4

Until 1853, there were many obstacles standing in the way of completing the sections of the Eastern Railway lines on either side of the Vistula River. During the bridge construction, the connections between Dirschau and Marienburg were very poor, especially in winter. An express service company in Königsberg had the task of transporting freight between Prussia, Pomerania and Brandenburg, as they had done in the past, providing a vital service in winter. The company took up accommodation opposite the Ostbahn to maintain certain delivery times but could not meet their obligations in bad weather and icy conditions. The company did not seem too eager to do business, either. As the newspapers reported at the time, shipments of railway goods took three days or more

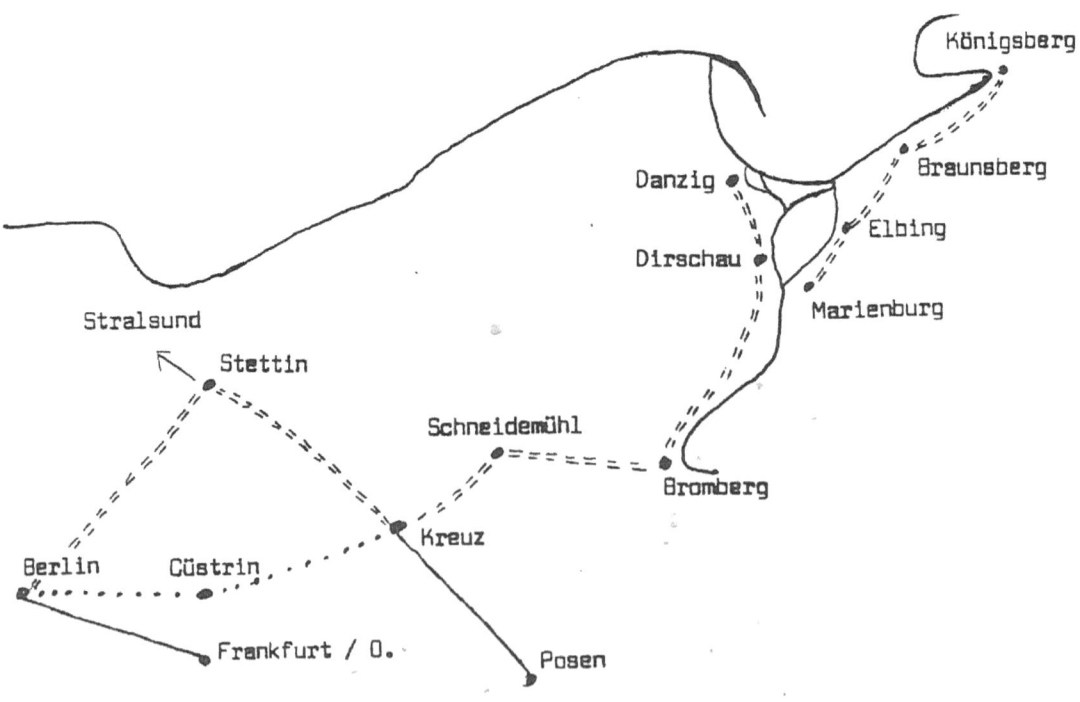

The proposed route connections for the Ostbahn in 1851

between Dirschau and Marienburg. In winter, goods often remained on the shelf for as long as they were unable to be transported over the ice.

Apart from the Dirschau-Marienburg section with the construction work on bridges, a direct connection to Berlin was delayed due to engineering costs and bad weather. Financial difficulties also delayed the completion of the direct connection between Kreuz and Berlin to the west. The trains continued to travel west via Stargard and Stettin.

Although the preparatory work for the Eastern Railway provided a direct route from Kreuz via Cüstrin to Berlin, these plans were changed, due to a lack of funds, until the route from Kreuz via Cüstrin to Frankfurt (Oder) had been completed.

Eastern Railway Connections at Nogat and Vistula Bridges 5

On 12 October 1857, the Kreuz-Cüstrin-Frankfurt (Oder) route opened, with connections to the Driesen, Friedenberg, Zantoch, Landsberg (Warthe), Vietz, Cüstrin, Podelzig and Lebus stations and stops at Alt-Carbe, Gurkow, Dühringshof, Döllens-Radung and Tamsel.

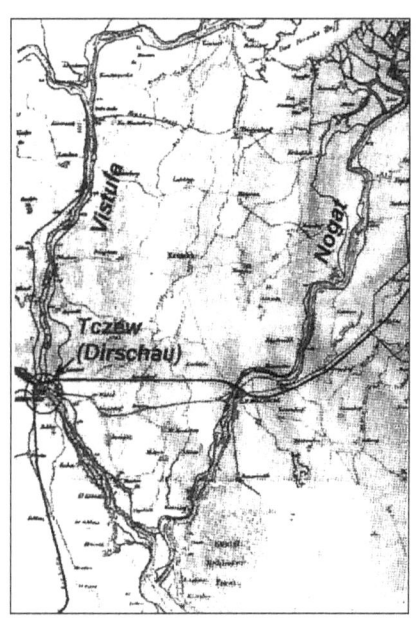

Map of the Vistula Delta with the Route of the Prussian Eastern Railway

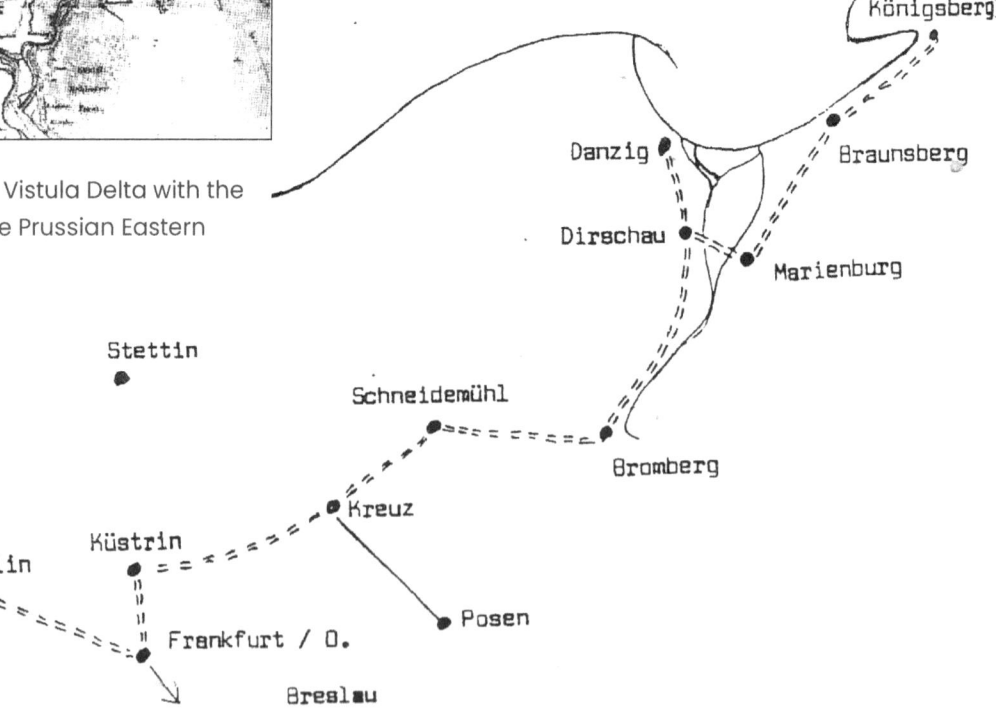

The proposed route connections for the Ostbahn in 1857

Following the completion of the bridges across the Nogat and Vistula Rivers, the bridges opened for traffic on 12 October 1857. The construction of bridges across rivers was rightly considered an engineering breakthrough at the time. A single-rail line lay in the middle, with a lane for horse-drawn trams on each side. Following the new route openings and improved connections, which were considered a "technical miracle" in the 19th century, a law of rights was published on 12 October 1857.

Since the three lanes were not separated from each other, and the distance between the two lattice girders was only 6.433 m, the bridge had to be closed to horse-drawn wagons while the trains were crossing. Outside the bars, narrow footpaths ran along both sides. A grand celebration was planned for the opening of the Vistula and Nogat bridges, but an outbreak of cholera in East and West Prussia caused the celebrations to be postponed. (Speculation was that it was the death of the king on 23 October 1857 and not the cholera that caused the postponement. His brother, Prince Wilhelm of Prussia, was conferred as the new regent on 7 October 1858.) A new Dirschau railway bridge was built between 1888 and 1890 due to increased demand.

Copy of an Early Railway Timetable for 1855

6

Berlin-Stettin-Posen-Bromberg-Danzig-Königsberg in Pr.

92 Miles in 23¼ or 19 hours. To Stettin Private, from the state railway. 30 pounds of luggage free.

Meilen	Personengeld I. Cl. Th. Sgr.	Personengeld II. Cl. Th. Sgr.	Personengeld III. Cl. Th. Sgr.	Stationen	Personenzüge 1 Früh	2 Früh	3 Mitt.	4 Abd	Schn- zug. 5 Abd	Güt.- zug. 6 Früh	7	Haupt-Anschluss-Posten (S. auch folg. Seite)
				aus Berlin		6 15	12 15	5 30	10 40	8 30		
	1 10	1 0	0 20	Neustadt-Eberswalde		7 32	1 32	6 48	11 45			
6	Von Stettin ab											
9,4	2 0	1 15	1 0	Angermünde		8 14	2 19	7 31	12 23			
11,9	2 20	2 0	1 10	Passow		8 46	2 50	8 2	12 52			
17,9	4 0	3 0	2 0	in Stettin		9 47	3 52	9 10	1 47	1.57		
				aus Stettin		10 8	4 19	7 10	2 2			
22,4	5 0	3 21	2 16	Stargard		11 24	6 10	9 17	3 0			
27,0	5 29	4 11	3 2	Aruswalde		12 19	7 25	10 32	3 86			
31,2	6 27	5 0	3 17	in resp. aus Woldenberg		1 12	9 18	11 53	4 27			
34,2	7 17	5 14	3 28	in Kreuz*		2 6	9 58	12 48	5 2			
				aus Kreuz		2 22			5.14			
45	9 27	7 2	5 5	in Posen		4 40	mit Güterzug III. Classe		8 1			
				aus Posen		11 b.			8 38			
				in Kreuz		1 53			10 50			
				aus Kreuz		5 40	2 8		5 2			
42,0	9 7	6 19	4 25	Schneidemühl		7 59	3 40		6 16			
50,0	10 29	7 25	5 23	Nakel		10 49	5 18		7 38			
53,6	11 22	8 11	6 5	in Bromberg		11 42	5 58		8 18			
				aus Bromberg		12 7	6 5		8 18			
59,6	12 27	9 5	6 24	Terespol		1 37	7 11		9 10			
62,6	13 21	9 21	7 7	Warlubin		2 41	7 56	Von Stettin ab Güterzug mit Personen II. u. III. Cl.		9 53			
65,1	14 7	10 3	7 15	Czerwinsk		3 23	8 25		10 17			
67,8	14 25	10 15	7 25	Pelplin		4 3	8 58		10 43			
70,5	15 12	10 27	8 4	in Dirschau*		4 53	9 40		11 15			
				aus Dirschau		5 56	9 48	7 fr.	11 22	11 22		
74,7	16 9	11 16	8 19	in Danzig		7 0	10 40	7 53	11 50	11 56		
				aus Danzig		5 35	8 20	4 km. 10 0	10 0			
				in Dirschau		6 29	9 14	4 54	10 58	10 58		
				aus Dirschau (Post)		7 4	9 56			11 37		
72,7	16 4	11 19	8 21½	Marienburg		10 34	1 30			3 .7		
76,6	17 0	12 7	9 5½	Elbing		11 52	2 27			4 0		
84,0	18 17	13 10	10 1½	in Braunsberg		2 28	3 13			5 36		
92,2	20 10	14 16	10 29½	Königsberg in Pr.		5 13 Abd.	6 7 Früh			7 15 Abd.		

Express train between Berlin and Dirschau with passengers I. and II. Class. Between Königsburg and Marienburg all trains with coaches I., II. and III class.

Neisse-Brieg.

No. 54. 6½ Meilen in 1¾ Stunden. Privatbahn. 50 Pfund Gepäck frei.

Meilen	Personengeld I. Cl. Th. Sgr.	II. Cl. Th. Sgr.	III. Cl. Th. Sgr.	Stationen	Personenzüge 1 Mrgs.	2 Vormt.	3 Abds.	4
	Von Neisse ab			aus Neisse	6 10	10 0	6 10	
,1	0 10	0 7	0 4	Bösdorf	6 25	10 15	6 25	
3,4	0 20	0 15	0 10	Grottkau	7 4	10 49	7 4	
6,2	1 10	1 0	0 20	in Brieg	7 40	11 19	7 40	

Haupt-Anschluss-Post von Neisse nach Glaz, 7 Meilen, täglich 10½ Uhr Vormittags in 7½ Stunden für 1 Thaler 12 Silbergroschen.

Rail Time Tables are available from the Museum of Communication, Frankfurt

Completion of Eastern Railway Connections from West to East Prussia

7

Following the completion of the rail line from Marienburg to Braunsberg on 19 October 1852, the route was transferred to the Ostbahn management to serve as a regular traffic route.

On 2 August 1853, provisionally the last stretch of the Eastern Railway from Braunsberg to Königsberg was put into operation. A snowy winter and the cholera in 1852 had delayed the completion. The station in Königsberg was highly praised for its furnishings and construction. It is said to have been the largest and most magnificent station building in Prussia and Germany.

The towns of Heiligenbeil, Wolittnik, Ludwigsort, Kobbelbude and Seepothen were connected to the Eastern Railway as the line approached Königsberg. At that stage, only the 18 km connection between Dirschau and Marienburg over the Vistula and Nogat bridges was still to be established. The completion of the bridges was delayed due to unforeseen geographical obstacles.

View of the Königsberg station, Prussia, 1 January 1854 (German Reich's Bank Calendar 1935)

As of now only the 18 km long connection between Dirschau and Marienburg over the Vistula and Nogat bridges was to be established. The completion of the bridges faced unforeseen geographical obstacles leading to delays.

The inaugural train entered Königsberg station on 2 August 1853. The hall was 126 m long and 33 m wide. It consisted of one central platform and two side platforms, whilst on the west side of the station, there were two open platforms. It remained in service until 1889, with trains continuing to Labiau and Tilsit. A newspaper of the day reported: "A new era has begun when Königsberg was connected to the German railway network by the Eastern Railway." The inauguration of the station was a veritable carnival, and a pageant was presented on the platform. When the railway line continued eastwards, it reached **Eydtkuhnen** on the Russian border in 1860 and connected to the Russian network. Grain imports increased greatly because Königsberg was the terminus of the shortest connection between the Black Sea and the Baltic Sea.

Further Development on the Eastern Railway

8

Traffic on the Eastern Railway lines developed unusually well. According to information available from the archives, from the beginning of 1864, and on several sections of the second route, passenger traffic increased as people thrived on the fast-developing opportunities.

The train service from Cüstrin via Frankfurt (Oder) to Berlin was in urgent need of expansion. The land needed was acquired before the law of May 1865 was passed, and approval for the construction of the Cüstrin-Berlin double-track line was granted. The section from Frankfurt (Oder) to Berlin was completed ahead of the rest of the route because of the increased volume of traffic flowing from the Lower Silesian (Niederschlesisch-Märkischen) lines, which benefited from the connection between Frankfurt (Oder) and Berlin. The extension of the Ostbahn to St.Petersburg and the Thorn-Warsaw line also contributed to the increase in traffic.

The Post-Expedition 34 had been located on the Ostbahnhof since 1 October 1867. It was renamed the Post Office on 1 January 1872, and on 1 July 1886, the Post Office moved to 102 Frankfurter Allee.

With the direct route from Cüstrin to Berlin, it was decided to build a special train station to relieve the Lower Silesian (Niederschlesisch-Märkischen) train station where the Ostbahn trains arrived. In 1866, the Prussian Minister of Commerce approved the widening of the gates of the city wall to accommodate the Frankfurt (Oder) railway so the extra rail lines could be incorporated into the Eastern Railway.

The first section of the railway line passing through the stations of Gusow, Golzow and Cüstrin was opened on 1 October 1866. The remaining distance to Berlin, passing through the Neuenhagen, Straußberg, Dahmsdorf-Müncheberg and Trebnitz stations, was completed on 1 July 1867. Due to delays in finishing the new Ostbahnhof in Berlin, the planned opening date of 1 July 1867 was never achieved. The minister in charge requested the postponement of the opening date for the whole line. On 1 October 1867, the entire Berlin-Cüstrin route and the East station in Berlin were opened to traffic.

THE EXPANSION OF THE EASTERN RAILWAY'S NETWORK ALONG TWO MAIN LINES FROM 1871 TO 1873

1. The map below shows the Thorn-Insterburg railway, running parallel to the route to Königsberg. It was opened on 15 August 1873, and on the same day, the line across the Vistula Bridge at Thorn was inaugurated. Until the completion of the bridge, freight wagons were transported over the Vistula on wide-beamed flat-bottom sailing platforms and pulled up with a wire rope onto the bank.

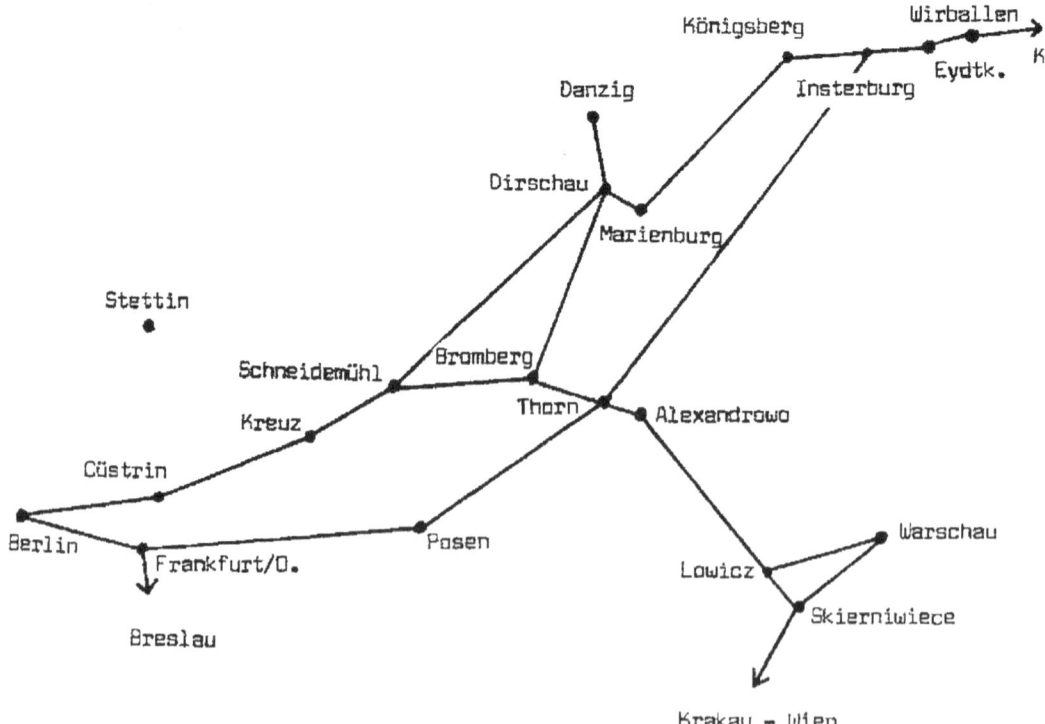

2. The Schneidemühl-Dirschau route was the shortest route to Königsberg by 36 km. It was connected by additional rail lines to Danzig in 1875/76 and finally received recognition as the main traffic route to the east – much to the disadvantage of Bromberg, which now sat between the two parallel lines Berlin-Dirschau-Königsberg and Berlin-Posen-Thorn-Insterburg. (See previous map.)

The main railway lines were joined during the time of the Franco-Prussian Wars, enabling faster troop movements east to west. In addition to some important private railway lines and other state railway lines, continuous branch lines were created, which enabled rural agricultural villages to move goods and services to important towns and cities in East Prussia. The cross-connections of the three major lines, Berlin-Danzig, Berlin-Königsberg and Posen-Insterburg, continued an uninterrupted service until the end of World War I. They faced changes following the signing of the Treaty of Versailles that ended World War I. It went into effect on 10 January 1920, resulting in redrawn international borders.

Eastern Railway Network between 1875-1876

9

The operational management of the Eastern Railway was headed by three senior officials, a superintendent inspector, a master technician and a managing director for marketing.

On 1 October 1873, the organisational management of private railways, due to increasing financial difficulties, passed to the Prussian State Railways Administration. The day-to-day commercial transactions carried out by the superintendents were also gradually transferred to newly created railway commissions.

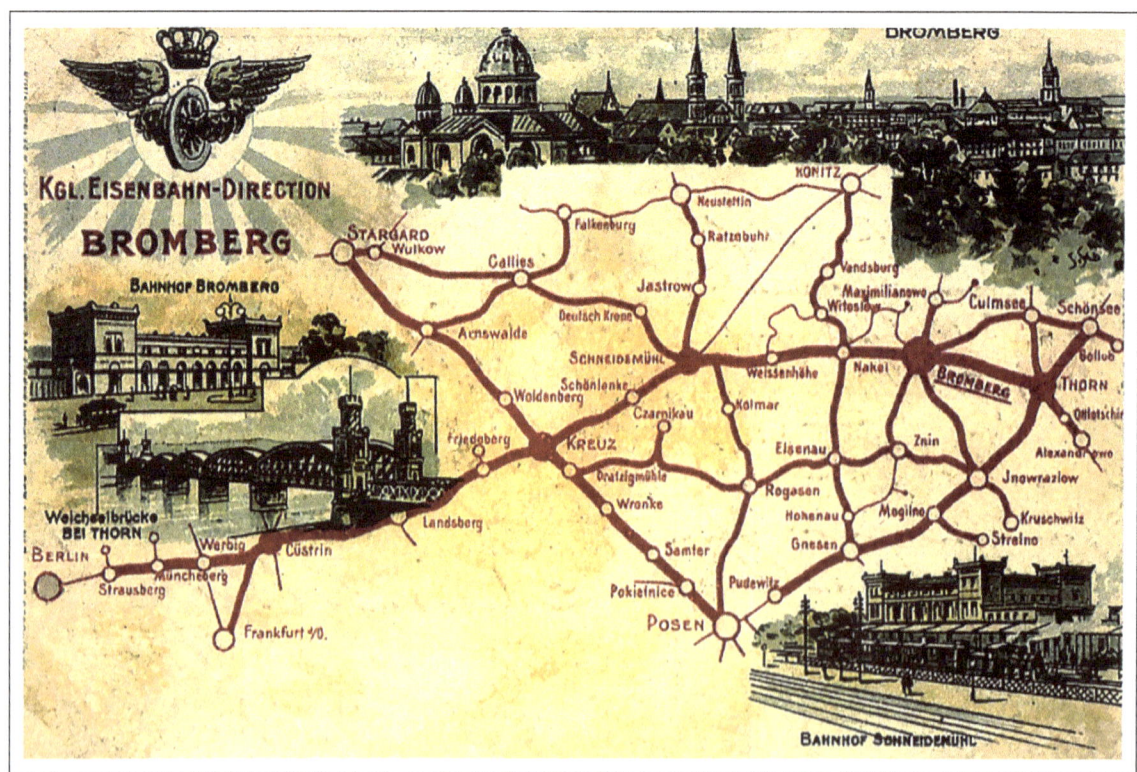

"Royal Railway Management Bromberg" The Centenary cover.

They were: Berlin (activity from 1.10.1873)
Königsberg (from 1.10.1873)
Bromberg (from 1.10.1874)
Danzig (from 1.1.1876)
Thorn (from 1.10.1876)
Schneidemühl (from 1.4.1877)
Stolp i.Pom. (from 1.4.1879)

The private Prussian railway company Berlin-Stettin (BStE), which existed from 1840 to 1885, was purchased by the Prussian state in 1877. A special railway commission in Stettin was established on 24 December 1877 to oversee functions.

By the supreme announcement of 24 November 1879, the railway commissions were replaced by Royal railway depots in Berlin, Königsberg, Bromberg, Danzig, Thorn, Schneidemühl and Stolp i.Pom. A special office in Stettin managed the Pomeranian Railway and improved connectivity to agricultural villages. A change in the direction of the Eastern Railway by transfer or removal of routes did not impact the work of the Stettin management. From 1 April 1880, "The Royal Management of the Eastern Railway" changed its title to "The Royal Railway Management of Bromberg".

Frontier Stations and the Border Routes of the Eastern Railway

10

With the first part of the line opened in 1851, the construction of the Eastern Railway continued from Königsberg to the Russian border, and an extension was planned from Thorn to the Polish border and on to Lowicz. Both projects ran side by side.

The route to the Russian border opened up the possibility of connecting the Russian Railway through a connection from St. Petersburg to Warsaw most conveniently and economically, or it could be permitted to run mail services just beyond the Prussian border.

During the reign of Emperor Nicholas I (until 2 March 1855), only a few rail routes were built in Russia. The town of Zarskoje (Carskoje) Selo had been the site of the summer residence of the Imperial Family and visiting nobility since 1838. It was located 24 kilometres south of the centre of St. Petersburg. The emperor permitted the construction of the Petersburg-Moscow rail line, which was opened in 1851, so he could reach his summer residence. The Warsaw-Krakow Railway, which had been in operation since 1848, opened up the possibility of a further rail connection. A railway connection between St. Petersburg and Vienna, using the existing Warsaw-Krakow route, was agreed upon. It was to run from St. Petersburg via Dünarburg and Vilnius to Grodno and Bialystok, providing a straightforward connection via Warsaw. This route was completed by the Prussian state. (See map in Chapter 13.)

On the Prussian side, the aim was to encourage the Russians to build the Russian railway from Dünaburg as close as possible to the Prussian border – not via Vilna, but more to the west via Kowno. The Russian government did not comply with this plan, as it would have created a straight line from Petersburg to Warsaw, but it approved a branch line from Vilna via Kowno to Eydtkuhnen.

On 7 May 1855, the Prussian Minister of Trade was authorised to communicate with the Russians and agree to special conditions for the passage and connection of the railways. On this occasion, he was to advocate the continuation of the Eastern Railway to Poland, from Bromberg via Thorn and Lowicz to Warsaw.

The negotiations were finalised in an international treaty, the final protocol of which was signed on 14 February 1857 by their respective delegations. (See reprint of the contract in Chapter 12.) Due to the differing gauges of the Russian and Prussian railways, it was argued by the Prussian side that the gauge should be the English gauge of 5 feet. Given the complexity of connectivity of rail lines across frontiers, the Prussian area supported a gauge of 4 feet 8½ inches, which remained unchanged for a considerable period of time in an area where the density of traffic increased following the connection to east Prussian towns. Due to the diversity of the rail lines on either side, the attempt to fulfil a public request to extend the Prussian rail lines or their branch lines to the Russian main line was unsuccessful.

In 1895, the state railway administration was reorganised and districts were partitioned. The Eastern Railway experienced its largest expansion as a result of war requirements, as well as the expansion of commerce. The network of the old Eastern Railway, which covered almost 5,000 km, was essentially allocated to the new railway managements in Bromberg, Danzig and Königsberg. This disappeared in 1919, at the end of the First World War, after almost 50 years of existence.

Earliest Known Copy of Prussian Railway Tariffs

11

In 1855, work on the Eastern Railways was progressing, and their railway connections in Central Europe were not yet complete. Courier tariffs were published in advance, before rail connections were completed in the district of Posen.

Staaten.	Extrapost-Taxe für 1 Pferd auf 1 Meile.	Courier-Taxe für 1 Pferd auf 1 Meile.	Wagengeld für 1 Meile.	Staaten.	Extrapost-Taxe für 1 Pferd auf 1 Meile.	Courier-Taxe für 1 Pferd auf 1 Meile.	Wagengeld für 1 Meile.
Anhalt-Bernburg	10 gGr.	14 gGr.	8 gGr.	Steiermark	35 kr.	45 kr.	17 kr,
„ Cöth. u. Dess.	10 Sgr.	15 Sgr.	7½ Sgr.	Kärnthen und Krain	36 „	46 „	17½,18kr.
Baden	48 kr.	. .	25 kr.	Tyrol u. Vorarlberg	40 „	47 „	18½ kr.
Baiern	44 „	. .	24 „	Siebenbürgen	34 „	. .	15 „
In den Hauptstädten	52 „	. .	. .	Küstenland (Triest)	36 „	. .	18 „
Belgien (per Post)	1 Fr. 50 C.	. .	1 Fr. 50 C.	Lombardei und Venedig per Post	3 Lr. 60 C.	4 Lr. 60 C.	1 Lr. 80 C.
Braunschweig	9 gGr.	13 gGr.	6 gGr.				
Bremen	27 Groote	36 Groote	24 Groote	Oldenburg	27 Groote	36 Groote	20 Groote.
Dänemark	44 Rb.sch.	. .	16 Rbsch.	zu Eutin	16 Sch.	24 Sch.	4 Sch.
Frankfurt a. M.	1 fl.	1 fl. 15 kr.	30 kr.	z. Birkenf. u. Oberst.	12½ Sgr.	17½ Sgr.	7½ Sgr.
Frankreich	2 F. p.Myr	. .	2 F. p. myr	Polen (pr. Werst)	5 Kop. S.	7 Kop. S.	2½ Kp., in incl. Postillonstrinkgeld. Warschau 5.
Hamburg	1 Mk. 4 Sh	2 Mark	8 Sch.				
Hannover	9 gGr.	12 gGr.	8 gGr.	Preussen:			
Hessen (Kurf.)	12½ Sgr.	17½ Sgr.	5 Sgr.	in Westphalen und Rheinpreussen	12½ Sgr.	17½ Sgr.	7½ Sgr.
zu Cassel	15 „	20 Sgr.	5 „	Übrigen Provinzen	10 Sgr.	15 Sgr.	7½ „
Hessen (Grossh.)	45 kr.	52½ kr.	30 kr.	Reuss ält. u. jüng. Lin.	11½ Sgr.	16½ Sgr.	7½ „
zu Mainz	52½ kr.	1 fl.	30 „	Russland (per Werst)	2½ Kp. S.	. .	12 Kp. S.
Hessen-Homburg	52½ „	1 fl. 7½ kr	30 „	in Finnland	4 Kop. S.	. .	p. Station
Hohenzollern	45 „	. .	15 „	Sachsen (Königreich)	10 Ngr.	15 Ngr.	5 Ngr.
Holstein	51 Rbsch.	77 Rbsch.	26 Rbsch.	Sachsen-Weimar	11½ Sgr.	16½ Sgr.	7½ Sgr.
Lauenburg	51 „	77 „	26 „	Sachsen-Meiningen	42 kr.	1 fl.	30 kr.
Lippe-Detmold	10 gGr.	14 gGr.	6 gGr.	in Liebenstein in 8.	44 „		
Lippe-Schaumburg	12½ Sgr.	17½ gGr.	5 Sgr.	Sachsen-Coburg.-Gotha	42k.11/Sg	1fl.16½Sg	30 k 7½ S.
Lübeck	1 Mk. 2 Sh	. .	12 Sch.	Sachsen-Altenburg	10 Ngr.	15 Sgr.	5 Sgr.
Luxemburg (p. Post)	1 Fr. 75 C.	3½ F.p.P.	1 F. p. Pst	Schleswig	51 Rbsch.	77 Rbsch.	26 Rbsch.
Mecklenb.-Schwerin	20 Sch.	28 Sch.	12 Sch.	Schwarzbg.-Rudolst.	11½ Sgr.	16½ Sg.42k	7½ S 30 k.
„ Strelitz	12½ Sgr.	17½ Sgr.	7½ Sgr.	Schwarzb.-Sondersh.	11½ g.42k	16½ Sg.2fl	7½ „, 30 „,
Nassau	45 kr.	1 fl.	30 kr.	Schweiz:			
Niederlande	1 fl. p. Pst.	1 fl. p. Pst.	1 fl. p. Pst.	pr. poste	4 Fr.	. .	2 Fres.
Oesterreich:				1 Post = 3 Schwz. Stunden.			
ob u. unter d. Enns	35, 37 kr.	45, 47 kr,	16½-18kr				
Salzburg	36 kr.	46 kr.	17 kr.	Waldeck	12½ Sgr.	17½ Sgr.	7½ Sgr.
Böhmen, Mähren	36 „	46 „	17 kr.	Württemberg	45 kr.	. .	30 kr.
Schlesien u. Galizien	36, 30 kr.	46, 37½ kr	17, 15 kr.	in Stuttgart 15 kr. pr. Pferd mehr über die ganze Taxe.			
Ungarn	34, 35 „	. .	15, 17 „				

Postillione erhalten in der Regel statt des tarifmässigen Trinkgelds: Bei 2 Pferden auf 2 Meilen 20½ Sgr. oder 1 fl. 12 kr., auf 2½ Meilen 26 Sgr. oder 1 fl. 30 kr., auf 3 Meilen 1 Thlr. oder 1 fl. 48 kr.; bei 3 Pferden auf 2 Meilen 28 Sgr. oder 1 fl. 36 kr., auf 2½ Meilen 1 Thlr. 5 Sgr. oder 2 fl., auf 3 Meilen 1 Thlr. 12 Sgr. oder 2 fl. 24 kr.; bei 4 Pferden auf 2 Meilen 1 Thlr. 5 Sgr. oder 2 fl. auf 2½ Meilen 1 Thlr. 14 Sgr. oder 2 fl. 30 kr., auf 3 Meilen 1 Thlr. 22 Sgr. oder 3 fl.

Ueberdem kommen noch die Chaussee- und Brückengelder, circa 5 Sgr., und an Wagenmeistergebühren und Schmiergeld ca. 2½ — 5 Sgr., von Station zu Station zur Erhebung.

International Agreement between Russia and Prussia 12

On 6 June 1860, the Königsberg-Stallupönen line was opened, and on 15 August 1860, the final section from Stallupönen to Eydtkuhnen was complete. At this time, the exchange post office for Prussian-Russian Post was relocated from Tilsit to Eydtkuhnen.

In Russia, the St. Petersburg-Gatschina route was opened on 1 November 1853 and the Gatschina-Luga line on 5 December 1857. According to Russian communications, the opening of the Luga-Pskow line was imminent, and in the spring of 1858, work was to begin from Pskow, Russia, to Vilna and the Prussian border. On 7 February 1860, the 52 km section from Pskow to Dünaburg was opened, and on 3 November 1860, a connection through to Ostrow was added. The completion of the remainder from Kowno to Eydtkuhnen was promised in December 1860.

On 11 April 1861, when the Eastern Railway administration had completed all available border station connections mentioned above and their inclusion in the Prussian time tables was possible, direct rail traffic between Berlin and St. Petersburg, Russia, commenced.

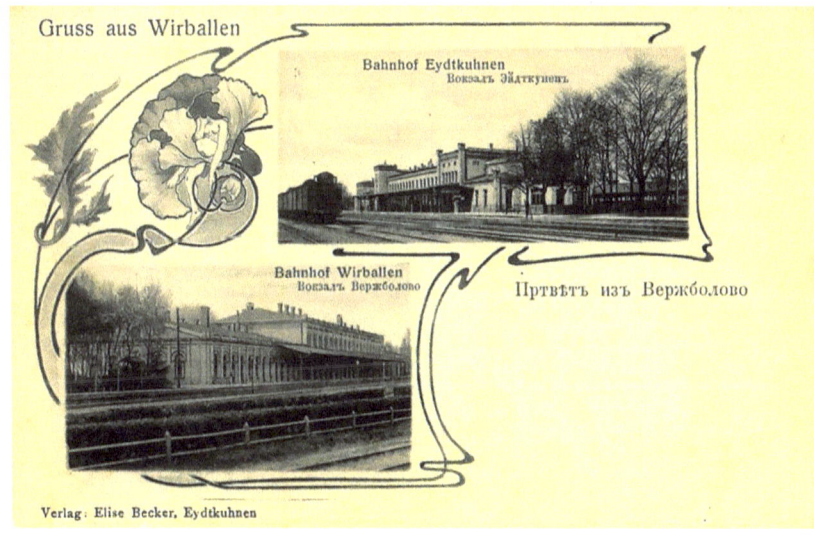

Unused postcard Gruss aus Wirballen 1906

On 1 October 1867, the Berlin-Cüstrin line in Pomerania, which had been a double track from the outset, was opened to traffic, together with the new Berlin reception building. With this, the direct line from Berlin to Cüstrin, Bromberg, Königsburg, Eydtkuhnen and St. Petersburg was completed. Completion of the line from Danzig to Neufahrwasser at the same time opened a network of direct rail connections with the Baltic Sea.

(Nr. 4700) Copy of the international agreement between Prussia and Russland, for the establishment of a continuous rail connection from Königsberg to St. Petersburg 14 February 1857.

Artikel 5.

Die Grunderwerbungen, die Erdarbeiten und die Kunstbauwerke sollen sogleich für ein Doppelgeleis bewirkt und ausgeführt werden; die Legung des zweiten Geleises kann aber bis dahin ausgesetzt bleiben, bis das Bedürfniß dazu von den betreffenden Regierungen anerkannt wird.

Artikel 6.

Die Bauarbeiten sollen soweit als thunlich dergestalt gefördert werden, daß die Preußische und die Russische Bahnstrecke zu gleicher Zeit vollendet und in Betrieb gesetzt wird.

Die Königlich Preußische Regierung soll indeß nicht verpflichtet sein, die Bahn von Königsberg nach Eydtkuhnen früher zu vollenden, als auch die Eisenbahn von St. Petersburg bis zu demjenigen Punkte, von welchem die Russische Zweigbahn von der Hauptbahn abgehen wird, vollendet sein wird.

Mit Rücksicht hierauf wird die Kaiserlich Russische Regierung die Königlich Preußische Regierung wenigstens zwei Jahre vorher davon benachrichtigen, bis zu welchem Zeitpunkte die Russische Zweigbahn und die Bahnstrecke von St. Petersburg bis zu dem Abgangspunkte der Zweigbahn von der Hauptbahn fertig sein wird.

Artikel 7.

Der Betriebswechsel soll an der Grenze stattfinden in der Weise, daß die Preußische und die Russische Eisenbahnverwaltung jede für sich einen besonderen Endbahnhof in unmittelbarer Nähe der Grenze auf ihrem Gebiete anlegen und die Preußischen Bahnzüge auf dem schmaleren Preußischen Geleise in den Russischen Bahnhof, die Russischen Züge auf dem breiteren Russischen Geleise in den Preußischen Bahnhof einfahren.

Artikel 8.

Der Bau, die Unterhaltung und die Beaufsichtigung des breiteren Geleises auf Preußischem Gebiete, zwischen dem Preußischen Endbahnhof und der Grenze, liegt der Preußischen Verwaltung ob; der Bau, die Unterhaltung und die Beaufsichtigung des schmaleren Geleises auf Russischem Gebiete, zwischen dem Russischen Endbahnhof und der Grenze, liegt der Russischen Verwaltung ob. Für das Befahren und die Benutzung dieser Theile der Eisenbahn sollen die Preußische und die Russische Verwaltung eine der andern keine Vergütung zu zahlen haben.

Die Brücke über die Lipona soll auf gemeinschaftliche Kosten, und zwar zu gleichen Theilen zwischen der Preußischen und der Russischen Verwaltung gebaut und unterhalten werden.

Artikel 9.

Die Hohen kontrahirenden Theile werden dafür sorgen, daß in den Endbahnhöfen die erforderlichen Einrichtungen getroffen werden, um mit dem möglichst geringen Zeit- und Kostenaufwande die durch den Unterschied der Spurweite bedingten Umladungen der Güterwagen bewirken zu können.

Artikel 10.

Für die Endbahnhöfe bei Eydtkuhnen soll zwischen den Verwaltungen der beiden Eisenbahnen, unter Genehmigung der betreffenden Landesbehörden, ein übereinstimmendes Reglement für die Signale und alle Einzelnheiten des Betriebes vereinbart werden.

Artikel 11.

Die beiden Eisenbahnverwaltungen werden Behufs der zweckentsprechenden Regelung des Fahrplans, besonders der durchgehenden Züge, unter Genehmigung der betreffenden Landesbehörden, sich verständigen.

Artikel 12.

Der Fahr- und Frachttarif wird von jeder der beiden Eisenbahnverwaltungen für ihr Gebiet festgesetzt und der anderen Verwaltung mitgetheilt werden.

Artikel 13.

Es soll sowohl in Betreff der Beförderungspreise, als der Zeit der Abfertigung, vorbehaltlich des durch die Zollvorschriften bedingten Aufenthalts, kein Unterschied zwischen den Bewohnern beider Staaten gemacht werden, wobei sich von selbst versteht, daß diese Zollvorschriften für die Bewohner beider Länder eine gleichmäßige Anwendung finden sollen.

Artikel 14.

Für den Fall, daß eine der beiden Regierungen es vorziehen sollte, sich nicht selbst mit dem Bau und dem Betrieb der Eisenbahn auf ihrem Gebiete zu befassen, sondern solche einer Privat-Gesellschaft zu überlassen, wird die betreffende Regierung darauf Bedacht nehmen, die pünktliche Ausführung der Bestimmungen der gegenwärtigen Uebereinkunft sicher zu stellen, und sich die geeignete Einwirkung auf den Betrieb vorzubehalten.

Artikel 15.

Alle polizeilichen und zollamtlichen Maaßregeln, zu welchen die Betriebs-Eröffnung der den Gegenstand der gegenwärtigen Uebereinkunft bildenden Eisenbahn Veranlassung geben sollte, bleiben einer jeden der beiden Regierungen vorbehalten und sollen, soweit als thunlich, vorgängig vereinbart werden.

In Betreff der Förmlichkeiten der zollamtlichen Revision und Abfertigung des Passagiergepäcks und der ein- oder ausgehenden Güter, sowie der Paßrevision, ertheilen beide Regierungen sich die Zusicherung, daß die Eisenbahn von Königsberg nach St. Petersburg nicht minder günstig, als irgend eine andere in das Ausland übergehende Eisenbahn behandelt werden, und daß im Interesse der Förderung des Verkehrs dabei jede nach den in beiden Staaten bestehenden Gesetzen zulässige Erleichterung und Vereinfachung stattfinden soll.

Artikel 16.

Vor der Betriebseröffnung der beiden Eisenbahnen werden die Regierungen sich in Betreff der Veränderungen, welche die neue Verbindung in dem Betrieb der Posten und Telegraphen herbeiführen könnte, näher benehmen.

Artikel 17.

In allen Fällen, wo die Eisenbahn-Verwaltungen des einen oder des anderen Staates über die verschiedenen in der gegenwärtigen Uebereinkunft vorgesehenen Punkte, und überhaupt über die, den Zusammenhang des Betriebes zwischen beiden Grenzen und das Gedeihen des Transithandels sichernden Mittel sich nicht sollten einigen können, werden die Regierungen von Amtswegen einschreiten und sich über alle zu ergreifenden Maaßregeln verständigen.

Artikel 18.

Die gegenwärtige Uebereinkunft soll ratifizirt und die Auswechselung der Ratifikations-Urkunden zu Berlin im Laufe eines Jahres, vom Tage der Unterzeichnung ab gerechnet, oder wenn thunlich früher bewirkt werden.

Zur Beglaubigung dessen haben die Bevollmächtigten dieselbe unterzeichnet und besiegelt.

So geschehen Berlin, den $\frac{14}{2}$ Februar 1857.

(L. S.) v. d. Reck.
(L. S.) Scheele.
(L. S.) Saint-Pierre.
(L. S.). de Kerbedz.

Continued from last page.

The Eastern Railway's Continuous Connections to the Russian Border 1862

13

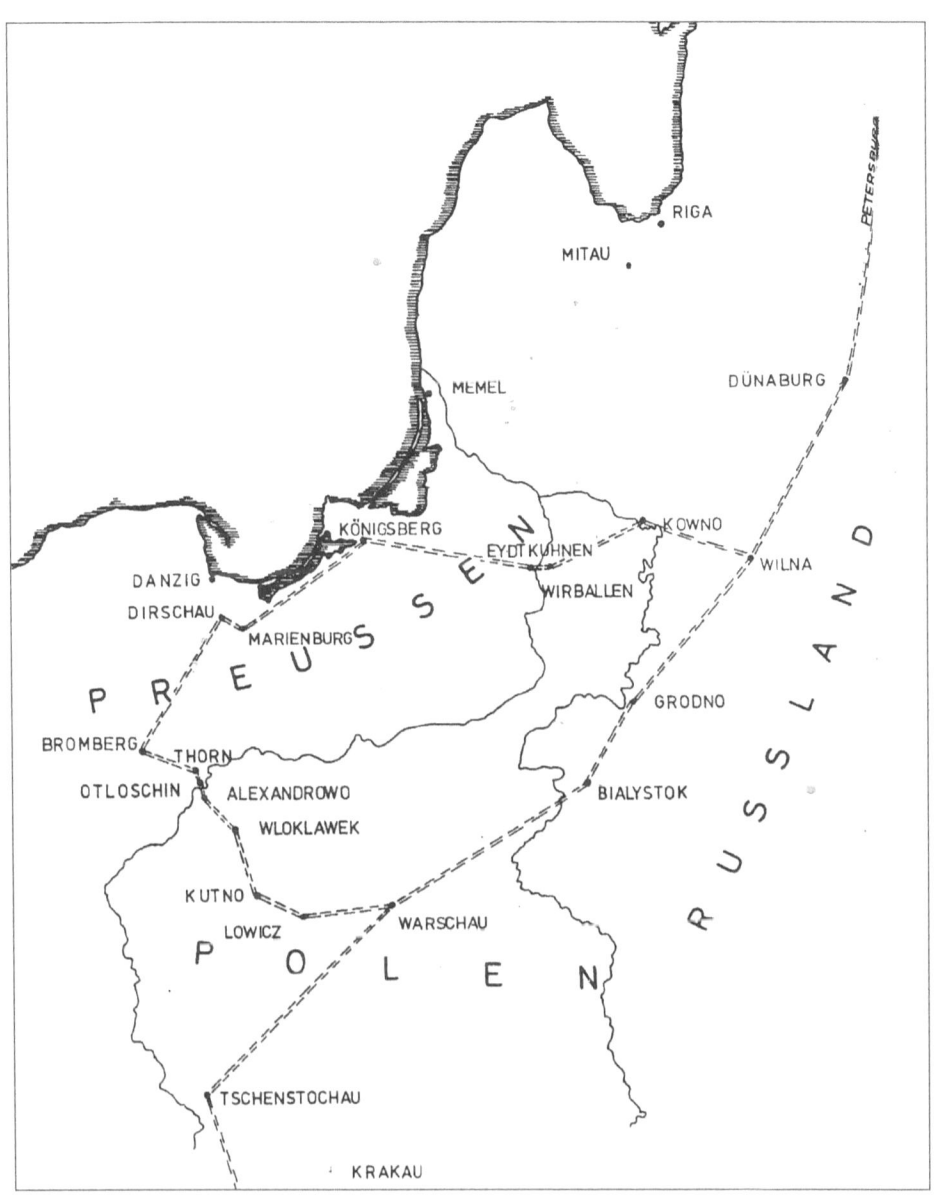

Proposed connections linking the Eastern Railway at Bromberg with Warsaw (Poland) and Wilna (Russia)

The Eastern Railway's Connections at the Polish Border

14

As well as the preparations for the border route from Königsberg via Eydtkuhnen to the Russian border, those from Bromberg via Thorn to the Polish border were also agreed upon.

Following the completion of connections and the continuation to Lowicz on the Polish side, one could expect a reasonably continuous connection to Warsaw. From 1845, Lowicz was connected by a rail line to Warsaw and the Warsaw-Vienna railway.

For Prussia, Thorn was not only an important frontier fortress; it was also intended to connect the Kingdom of Poland with the Baltic ports of Prussia through the connection to the border and on to Warsaw. In particular, direct rail connections along the ancient trade route from Poland to Danzig would help promote its development. Ideally, all traffic routes should reach Poland, Hamburg, Belgium, France and England.

The railway line was to run from Bromberg in an easterly direction on the left bank of the Brahe, cross the Brahe at Czersk (Brahnau), and continue along the left bank of the Vistula to Thorn, where the station was housed in the bridgehead of the Thorn fortress.

The municipal councillors of Thorn organised petitions on behalf of the City Council, drawing the attention of the state to the importance of connecting the railway to the city itself. To prevent the frequent damage to the old Thorn wooden bridge by snowdrift and fire, a bridge with a massive substructure over the Vistula was advocated. Although the construction was approved, the commission felt the cost of a fixed bridge and a railway station for the city was way too high.

From Thorn, the line was to run all the way to the Polish border town of Otloczyn. The water mill at Otloczyn would provide a useful connection to the Prussian network. In

Prussia, the Eastern Railway line was built at public expense, while the construction of the Polish route to Lowicz, which was three times longer, was left to private entrepreneurs.

The gauge on both sides of the border was the standard Prussian gauge of 4 feet 8½ inches (1.435m), which was also adopted by the Warsaw-Vienna Railway. The contract for the construction of this railway was completed on 15 December 1856 and ratified on 19 February 1857. On the Prussian side, construction work on the Bromberg-Thorn route began in 1860. Since the station in Bromberg required additional facilities after the diversion of the Thorn rail lines, it was subsequently rebuilt and enlarged. On 24 October 1861, the Bromberg-Thorn railway line was opened with border stations at Czersk, Poland (German: Czersk; Heiderode 1942-1945) and Schulitz (Polish: Solec Kujawski).

On the Polish side, the completion was delayed due to the construction of a bridge over the Bzura River at Lowicz and a bridge over the Oehnja. The iron superstructures were delivered by the Borsig company in Berlin as much as a year late. The Polish route went through the cities of Alexandrowo, Wloclawek (German: Leslau) and Kutno.

On 4 December 1862, without any ceremony, continuous traffic was recorded on the Bromberg-Warsaw Railway.

"Fahrpost"

15

The term "fahrpost" was a generic term first associated with postal history and refers to the mass carting of mail or bulky goods using special carriages or stagecoaches, better known as horse-drawn carriages. They were used by the Prussian Post Office until the mid-nineteenth century for the transport of mail as well as paying passengers. Coaches were usually clumsy and uncomfortable and stayed at the pickup stations for hours. The post offices were unable to maintain discipline and order among the passengers and carters. On bad roads or during harvest time, they often refused to fulfil their obligations, under all sorts of pretexts. They usually provided bad horses with unsuitable harnesses and drove without prior knowledge of the post office. By a decree of 30 April 1712, it

Postcard celebrating postal delivery by horse-drawn carriage

was determined "that transporting of mail by horses under the name 'Extra post' will henceforth be a sovereign institution, consequently an accessory to the postal system."

Freight transport by land was arranged by several freight wagons. They were large, stacked wagons, covered with canvas under which the goods were packed in straw. They were often seen with four or more horses under the command of a post rider. These riders took off early in the morning on their long journeys, and it would be six to eight hours before they reached their destination.

Following the unification wars by Prussian against Austria and its allies in 1866/67, the Thurn and Taxis Post (the traditional private postal enterprise since the 16th century) came to an end. When the Prussian legislation passed the Reichspostgesetz, the last private postal enterprise was forced to turn over its business to the Prussian Post Office. When the German Reich was established in 1871, it declared a special monopoly law regarding the conveying of letters and newspapers and extended it to all its territories. In January 1876, a Reich Post Office under Postmaster General von Stephan was split off from Bismarck's Reich Chancellery as a government agency in its own right.

The Prussian Travelling Post Office

16

EARLY HISTORY

The first rail lines of the early "steam train" were opened in the 1830s, e.g., Liverpool-Manchester on 15 September 1830, Nuremberg-Fürth on 7 December 1835, and Berlin-Potsdam on 29 October 1839. The postal authorities immediately recognised the new means of transport as a good way to have mail between certain locations regularly delivered at fixed times. A law was passed in Berlin in November 1838 that imposed on the railway companies in Prussia an obligation to transport mail (closed letters, parcels and cash) by rail as required and to transport the mail and the necessary personnel free of charge.

In 1839, the Prussian railway staff first trialled a postal transportation service on the railways between Berlin and Potsdam. Horse-drawn post wagons were driven from the countryside to the station, and the post wagon was lifted off the carriage by means of a lifting device on a "railway plateau".

Loading a coachman's box at the railhead 1844

A "postilion" accompanied the mail on its journey and carried out the exchange of mail (bags and parcels). The postilion had to sit in the windstorm on the coachman's seat. At the end of the ride, the carriage was lifted and lowered again, hitched up and driven to the post office. In 1846, this type of mail transport ended.

In 1851, the first three-axle railway post wagons, with a length of 25 feet (7.85 m), were in use. The accompanying staff were called "postal conductors". On 1 November 1851, the first railway mail van departed Berlin for Leipzig. Attempts with two-axle mail wagons in 1852 did not prove successful, and they gave up the post runs at a very early stage.

With the expanding rail network and the economic boom, the postal services also increased their trips. In order to relieve the Expedition Offices of the post office and achieve a faster turnaround on the trains, "post-forwarding offices" ("Post-Speditions-Bureaus") were set up, whose personnel processed the mail during the journey. This proved a success when introduced in England in 1838, in Belgium in 1841, in France in 1846 and in Baden (Heidelberg) on 1 April 1848. In Prussia, the sorting of mail in travelling post offices began on 1 May 1849, and at the same time, the first eight Postal Dispatch Offices ("Post-Speditions-Ämter") in charge of Travelling Post Offices (Post Speditions Bureaus) were founded.

Following the Prussian postal reforms in 1849, the Railway Post Offices were permitted to transport letters and parcels up to the value of £40.00 (today's currency value) from 1 May 1849. In Metropolitan Berlin alone, from 1 January 1851, four Postal Dispatch Offices ("Post-Speditions-Ämter") were located at Berlin stations (Nos. I-IV) and were answerable to the Berlin Court Post Office. Other offices were in Breslau (No. V), Halle (Saale) (No. VI), Magdeburg (No. VII) and Deutz (No. VIII) and were fully equipped as functioning offices. On 1 February 1852, No. IX followed in Dortmund and on 1 April 1852, No. X in Cologne. As the Eastern Railway line extended year after year, the railway offices kept opening until the line reached Königsburg.

In 1842, the Stettiner Bahnhof opened as the terminus of the railway line to the then Pomeranian city of Stettin, connecting Berlin with the Stettin sea port. In 1877, the Berlin Stettiner Bahnhof became the Berlin passenger train terminus of the Prussian Nordbahn (Prussian Northern Railway) to Stralsund via Neu-Strelitz and, a few years later, to Rostock.

c.1842 Early-period locomotive departing Berlin Nordbahnhof (formerly Stettiner Bahnhof)

RAILWAY POST OFFICE MANAGEMENT

From 1852, the management of the postal operations on the railway trains was responsible for the deployment of operational staff, for their well-being and for the economic use of the railway mail wagons. Local post offices managed rail postal operations on individual routes on the same principles. The numbering of the railway post offices showed them as independent special offices and had significance in correspondence and communication.

In 1856 the managers were given the name Eisenbahn-Post-Speditions-Ämter and in 1875 were renamed Bahnpost officers. At the end of the rail mail era in Germany, they called themselves rail mail departments and were attached to a local post office. In 1870, postal areas were numbered to 21, initially, within the North German Federal Post Office. With the increase in traffic, the areas increased to 33, plus four in Bavaria and two in Württemberg. In the Reich postal area, by 1920 they had fallen to 27, only to rise again to 29 by 1937, 19 of which were independent offices.

Cost of arms of the imperial Post Office on a post wagon of the narrow-gauge railway.

In the course of rail mail history, rail post offices had their headquarters (some for only a short time) in Aachen, Altona, Augsburg, Berlin, Breslau, Bromberg, Chemnitz, Danzig, Dirchau, Dortmund, Dresden, Eisenach, Insterburg, Kassel, Cologne, Cologne Deutz, Königsberg (Pr.), Konstanz, Leipzig, Magdeburg, Mannheim, Mainz, Marienburg (Westpr.), Metz, Munich, Münster (Westf), Nürenberg, Oberhausen, Posen, Rheine, Saarbrücken, Schwerin, Soest, Straßburg (Els), Stuttgart, Trier, Ulm and Würzburg.

FURNISHING THE RAILWAY STATION

The interior of the Travelling Post Office wagon was constantly changing, according to experience and need. As a rule, there was the letter and packing room, washing facility, toilet, wardrobe, distribution compartments, bag clamps, folding chairs, rescue equipment, emergency first aid boxes, letterboxes and bagging tables. There were also so-called protective compartments at the front, on one side for storing shipments and on the other side as a safety buffer in the event that equipment boxes were being carried. In addition, there were guide overviews and aids, route stamps, sealing pliers, seals, forms, wrapping paper, binding cords, ink pads, flashlights and emergency lighting.

The franking and cancellation of mail items in the Travelling Post Office mail car

Since the population accepted the Travelling Post Offices (Post-Speditions-Bureaus) as a means of transport and delivery of letters, mail wagons with slots on the sides to drop letters into were provided until 1848. In 1851, the First Route Postmark appeared in Prussia on the back of the letters as per regulations, but that did not serve to validate the postage stamps on the front of the cover.

The earliest route postmarks were single and double circle postmarks in 1851, then a three-line and four-line concentric circle. The changes to postal regulations in 1859 brought about the discontinuation of the four-concentric-circle cancellation.

Travelling Post Office staff received the trip number formats (e.g. IV, Z, ZUG or Zug) from their respective railway post office, as well as the designation of the start and endpoint of the route they travelled on, e.g. Bahnpost 19 Frankfurt (Main)-Basel Zug 476. Above the mailboxes, the rail mail cars had exchangeable route signs with the inscription "Post nach...", as well as wagon route signs to ensure correct and timely transfer by the railway.

From 1868 in the northern German postal district and from 1871 following the unification of Germany, the Reichspost railway postmarks were standardised throughout the Prussian districts.

Commonly used was the three-line postmark with departure and arrival location, as well as day, month and Zug/train number. In the beginning, there were often two typefaces on the postmark. For example, for the return trip, there was usually a second postmark with the stations in reverse order. Or there was only one postmark for the route, with an indication of the direction of travel (**T-tour** or **R-return**). Circular postmarks were also used in the postal areas of Baden and Saxony before standardisation. During the early

Early examples of railway postmarks

period, it was common practice for the officials in the mail van to handwrite the station stop when letters and other items were collected. Few places were known in later years to continue with the practice of postmarking mail with a single-line postmark; examples are elusive to collectors.

DESIGNATION OF THE RAILWAY POSTAL SERVICES

Whoever built the routes and train stations were required to conduct postal operations. A sorting carriage was attached to the back of the trains; often the last carriage or a closed compartment was for sorted mailbags. The Association of German Railway Administrations was founded in 1847 to create a degree of uniformity among the 60 private railway companies. From 1 March 1865, there were regulations for goods and passenger traffic, which then became law on 10 June 1870.

Railway postal services were differentiated according to the tasks they had to deal with: A = All rail mail, re-working of all mail items; B = railway may transport letters and valuables; P = Parcel rail mail, re-working of parcels; newspaper items were in red in the postal rate book.

The number of people working in railway posts was based on the scope of the work. It fluctuated from one to over 20 officials, determined by the railway post offices, which then regulated the service with prepared duty timetables and railway regulation guides.

FORMS USED BY THE RAILWAY MAIL SERVICE

These have been changed and simplified over the decades. The increase in mail traffic made it necessary to relax the security regulations that had been considered essential for an efficient operation. Before railway posts were set up, all mail in Prussia had to be entered individually on post slips and in "railway operations books". A new "trip direction/expedition mode" was introduced with the railway guards.

A distinction was made between ordinary letters and other items, i.e. registered letters, insured letters, advance mail, letters containing cash, and parcels. Mail items were no longer entered in a book. Postage advances continued to be settled with the post offices until the introduction of postage stamps in 1850, and in 1867 these were generally ordered. The increase in mail traffic made the need to simplify operations even more pronounced. Here are some examples of the progress made:

1868-1870   First attempt at maintaining a constant record of data for the collection of common parcels.

1871   Valuable bag closed with a seal; reworking of valuable items; specification of the recipient on parcels.

1872   Sale of postage stamps; since 1920, on branch routes and narrow railways, the introduction of parcel cards.

1874	Printed matter card introduced.
1875	Express post dispatched from rail post to rail post and, in 1880, between railway posts and local post offices.
1878	Reworking of ordinary and registered letters in Schaffner/Conductor Guard railway posts; in 1890, including those pertaining to local post offices.
1892	Discontinuation of the letter bundle counting and written card closings.
1894	Station names in official abbreviations on loading slips and other forms.
1897	Loss of acknowledgement of cargo items rejected by railway guards.
1900	Handover of ordinary and inconsistent parcels without sorting.
1903	Use of lead seals instead of lacquer seals.
1904	Introduction of sealing pliers, and fountain pens replace dip pens.
1908	Summary treatment of valuable parcels up to a value of 600M. In 1909 also for insured letters.
1910	Letter pouch without a card closed with a lacing; wallet pouch only with a seal.
1915	Rejection of bag mail without a baggage loading slip.
1916	Use of light yellow bags (valuable) and light red bags (express).
1918	Value limit for summary treatment increased to 1200M.
1922	The position of Schaffner/Conductor Guard abolished.
1926	Bundles of letters tied with lace eyelets; no crediting of additional charges.
1927	Insured letters were redirected to 11-man railway posts.
1928	Express mail no longer counted.

All the above changes made the sorting and delivery of mail more efficient over the following decades. I would like to point out a ruling of 27 May 1874 when station letters were introduced. They were clearly marked and given the fastest possible transport to the railway station, and the route was noted on the shipment. From 1 January 1901, they could be received directly on the rail mail car. By an order of June 1906, carriage by carriage, with the help of railway personnel, was also possible. England and Austria were the first known countries with this practice of exchanging station letters, which they had done since 1 February 1891.

Use of the Mailbox on the Railway Mail Cars 17

Soon after the establishment of travelling post offices in the Prussian Post Office area (1 May 1849), the corresponding public began to deliver letters to the dispatching officials in the railway post offices shortly before the trains left. Since this caused the handover business to suffer multiple disruptions, letterboxes were attached to the railway mail wagons in 1849, initially to those on the Berlin-Deutz route and the newly established Eastern Railway routes until 1952.

Railway mail is understood to mean the processing and sorting (technically, "reworking") of mail items while the journey is in progress. Rail mail cars had special postal facilities for this and had mailboxes on the outside. Following its inception, the public embraced the new concept of dropping mail directly into a mailbox on the side of the carriage, and year on year, the volume of mail transported increased.

During my research at the Museum of Communication, Frankfurt, I found the number of mail items placed in the travelling post office mailbox for the 10-day period under review was as follows. The total number of items was 282,937 items, which would represent about two per cent of the 10,185,732 postal items delivered to the Deutsche Reichs-ost administration for one year. Of this, the day delivery (from 7:00 a.m. to 7:00 p.m.) accounts for 6,819,444 pieces and the night delivery (from 7:00 p.m. to 7:00 a.m.) 3,366,288 pieces.

At the following transfer stations, the delivery of the mailboxes in the railway mail cars was particularly significant:

STATION	DAYTIME MAIL VOLUME	NIGHT TIME MAIL VOLUME	STATION	DAY TIME MAIL VOLUME	NIGHT TIME MAIL VOLUME
Stettin	17,492	5,270	Danzig	8,747	8,159
Königsberg i. Pr.	8,747	8,159	Cöln	6,386	4,607
Berlin	6,314	4,909	Breslau	6,022	1,832
Frankfurt a.Main	5,809	3,053	Leipzig	4,892	2,697
Hamburg	4,515	2,870	Lübeck	3,231	1,246
Magdeburg	2,784	1,899	Breman	2,485	1,352
Kattowitz	1,972	1,352	Kiel	1,939	40
Rostock	1,729	886	Mühlhausen i.E	1,697	52
Ratibor	1,593	106	Dresden	1,510	375
Düsseldorf	1,468	486	Oppeln	1,447	166
Rendsberg	1,439	72	Straßurg i.E.	1,399	92
Nordhausen	1,376	103	Harburg	1,292	1,173
Chemnitz	1,196	304	Wrietzen	1,169	240
Liegnitz	1,167	327	Döbeln	1,132	327
Oldenburg i.Grh	1,125	584	Alt-Münsterol	1,092	309
Ohlau	1,073	386	Halle a.S.	1,055	262
Emden	1,038	15	Geestemünden	1,010	821

During the ten-day period, not all stations received large volumes of mail. At some stations, there were fewer than 1,000 letters in the mailboxes on the railway mail car at the transfer stations.

RAILWAY POST OFFICE STAFF

In the rail mail service, one was served by civil servants of the lower, middle and upper classes. They required physical as well as mental stamina and had to cope with an irregular way of life. They had to be mentally sharp, quick to grasp situations and able to take charge. They equipped themselves with in-depth knowledge of postal geography and postal employee relations. As a rule, special training was required. They had to deal effectively with unusual incidents such as railway-related irregularities and even accidents. In addition to his salary, the civil servant received an allowance for his unsocial hours, based on his absence from home. At the destination of the outward journey, the so-called overlying location, there was special accommodation where staff members could wait for the return journey.

RAIL MAIL POSTMARKS

Postmarks were primarily used for the official recording of loading papers and for operational evidence. They were also used as postage and cancellation postmarks. The characteristics of the postmark were determined by its fixed components; variable details, such as train number (ZUG), date, lines or data line and dots below the data line, were considered other characteristics. The three-line type of postmark characterised the North German Confederation post and the Reichspost in the pre-1889 era. This type of postmark was available for outward and return journeys. The digits of the data line were inserted, and the place of posting was noted by hand next to the stamped postmark.

In 1883, the Reichspost decreed that it would introduce a small oval stamp. The train number and date were always on two lines. The larger long oval postmarks were introduced on a trial basis in 1908; they displayed the train number and date on one line. Rail postmarks were introduced when needed, not always everywhere at the same time. No back stamps were used with the oval postmarks. The journey there and back could be seen only from the train number, and the direction of travel could be determined with the help of the postal timetable.

There were a multitude of types of railway postmarks. Arches, circles and hoop postmarks were used in Prussia, Mecklenburg, and Wuerttemberg; circular stamps in Baden, Bayern and Thurn & Taxis; concentric postmarks in Prussia, Saxony, Baden and Schleswig-Holstein; rectangular stamps in Prussia, Saxony and Baden un Bavaria. These can be seen in this volume and subsequent issues of my series.

Emergence of Freight Mail Cars on the Eastern Railway 18

The volume of mail to be transported made it necessary to use special mail cars on some railway routes, especially when a journey was very long and had a large number of stations receiving and sending their mail on the same train. Since only a short stop at each station was a condition of the railway company, the mail items intended for each station had to be ready. A brief description of the rail car is given below. Its essential components were the iron base and the wooden box.

The wheels were about three feet in diameter and, except for the hub, were made entirely of forged iron. Every two spokes, which were fitted into the hub as in an automobile, consisted of one piece, with part of the rim lying between them; they were, therefore, almost in the shape of a triangle with a curved third side. An iron tyre 1½" thick and 4-5" wide was drawn around the rim and connected to the rim by iron rivets. On the inward-facing edge of this tyre was the wheel flange, which was about an inch deep and an inch thick; this was to prevent the wheel from jumping off the rails.

The iron axle, about three inches thick, was firmly and precisely connected to the wheel so that the axle and wheel turned at the same time; the part of the axle that extended beyond the hub, the wart, which had a somewhat smaller diameter, rotated in bronze bushes which were enclosed by the iron sleeve. Finally, the axle and the spring above it were held together by iron straps and thus bore the entire weight of the box. [There is an opening in the surface of the can through which, in the case of the constant friction between the wart and the axle bearing, the heating caused by a mixture of tallow, palm oil, soda and water is brought into the can.]

On the springs lay two long wooden beams, connected by transverse wooden beams and covered with boards, forming the base of the body of the car. To reduce the bumps when several wagons followed one another, there were thick, hair-padded and leather-covered discs on the front ends of the long beams, each of which acted on a spring by means of an iron rod that went through the long beam. The ends of these springs lay horizontally under the car body and ended in the long beams; the ends passing through the centre to the outside served as a tie rod to reduce the bumps that occurred when braking.

The rail car itself was 16 feet long, 8 feet wide and 6 feet high. It had three doors on each side, which permitted easy loading and unloading from the three inner compartments. (See sketch below.) The middle section was mainly for the use of the conductor, which is why there was an upholstered armchair, a desk and a drawer for storing money boxes. From this compartment, sliding doors led into the other two compartments. In one of these, there were several partitions, one above the other, for stacking the letter bags, etc. These wagons could safely carry a load of 70 to 80 hundredweight and carried between 5 and 600 items. The maintenance was minimal, and with normal wear and tear, travelling costs were a maximum of six pfennigs per mile.

The second figure shows a six-wheeled mail wagon that was 25 feet long, 8 feet wide and 6 feet high. It offered significant packing space (1200 cubic feet). The internal divisions were the same as above.

In Halle an der Saale was Mr Winkens' factory for manufacturing railway wagons. His wagons were characterised by elegance, durability and economy and were used on the Berlin-Cölner route. On request, they were built so that the middle pair of wheels did

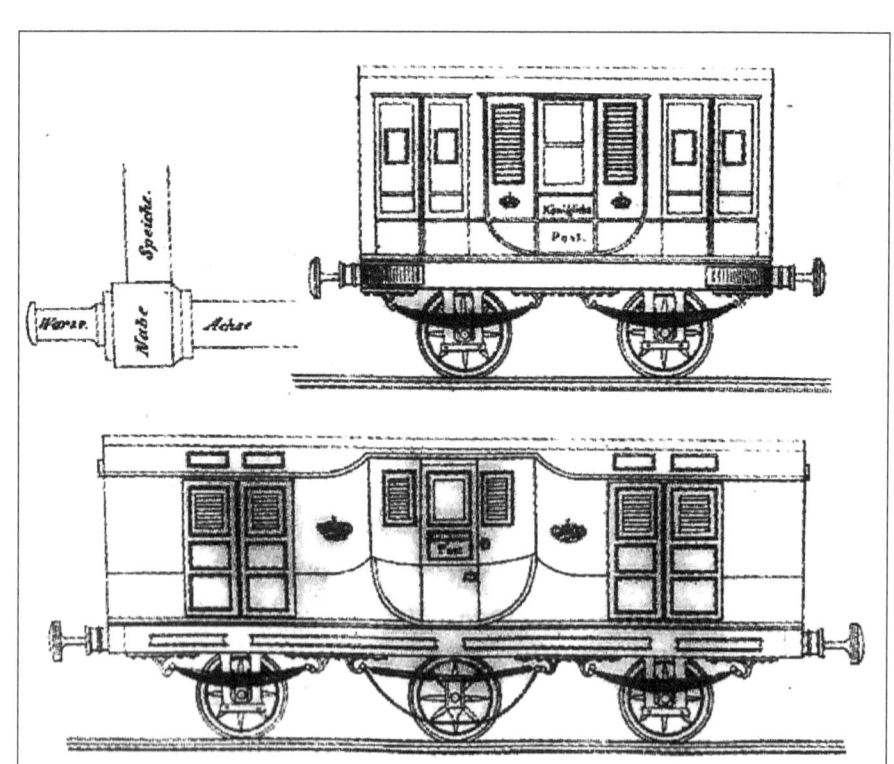

Depiction of the first German "Railway-Post Goods Wagon" in around 1840. Illustration in the German Postal Records from 1845. The design was based on the wagons of the stagecoach era.

not carry the usual leaf springs but were provided with a bow spring; this was formed from one leaf but consisted of two parts that met and fastened under the bush. The load capacity of these wagons was around 100 pounds.

The following was photocopied from the Deutsche Post Archives, page 137 of No. 11, 1873. The orthography of that time was adopted unchanged.

German Post-Almanac for the Year 1845

On frequent visits to the Frankfurt Post Museum in search of the earliest possible use of rail mail cars, I came across the yearbooks called the Deutscher Post-Almanac, published by Carl Herzog von Braunschweig. The contents were written by Postal Secretary Wilhelm Görges. Even as a rail mail driver, I was not allowed to copy them, but I was able to take photos of the relevant pages.

Gustav Wilhelm Emil Görges was a German postal worker and writer who began his publications around 1842. I found interesting texts and images from the years 1845, 1847, 1849 and 1851. In the volume from 1845, he described the use of rail freight cars in 1841/42. The literature on the earliest transport of mail by rail seems to be contradictory; a team of researchers found inaccuracies. It can be assumed that before 1840, there were regular railway mail wagons with two or three axles. They were in use on a small network of long-distance routes in 1938/39.

THE RAIL MAIL CAR

The rail mail car became the rolling duty station of the rail mail official. Tradition has it that in the early days of the railways, officials found the experience of working in a rail mail car like being on a lurching ship and it took some getting used to. When one coach carriage came off with the postilion, the Prussian railways designed a purpose-built carriage.

In 1841/42, the first railway mail wagons with three and two axles were introduced. In 1851, the Prussian Postal Administration owned ninety-two mail cars. Gradual changes to their interior made the work environment a joy. The staff benefitted from developments in technology and a workplace design adapted to their needs. Here are a few examples:

Four-axle compartment mail car of "Imperial Railway Management" (KED) Bromberg

1842	Two-axle, 5 m wagons were built and continued in use until 1851. They did not prove economically viable for the increased loads, and construction was later discontinued.
1883	Wagon width increased to 2.80m; new 5-axle 12m "Travelling Post Office" (BpW) wagon
1890-1892	New 4-axle 12m BpW for express trains
1898	2-axle BpW
1903	5-axle BpW; express train service discontinued
1907-1908	New 4-axle, 17m BpW; both ends of the car had a protective compartment, as well as a toilet, washing facilities and a wardrobe
1908	Special parcel wagons, 8.50m and 10m in length
1914	The Reichspost owned 2371 BpW and about 2500 post compartments in baggage cars
1925	All-steel D-Zug "Travelling Post Office" 20m in length
1927	Rail mail refrigerated trucks in use
1936	Letters and all mail BpW, 21.6m, fully welded, barrel roof
1938	15m "Travelling Post Office" wagon for express and passenger trains
1940	Total stock of "Travelling Post Office Wagons" was over 4177

Eastern Railway Main lines between Berlin and East Prussia 1851-1889

19

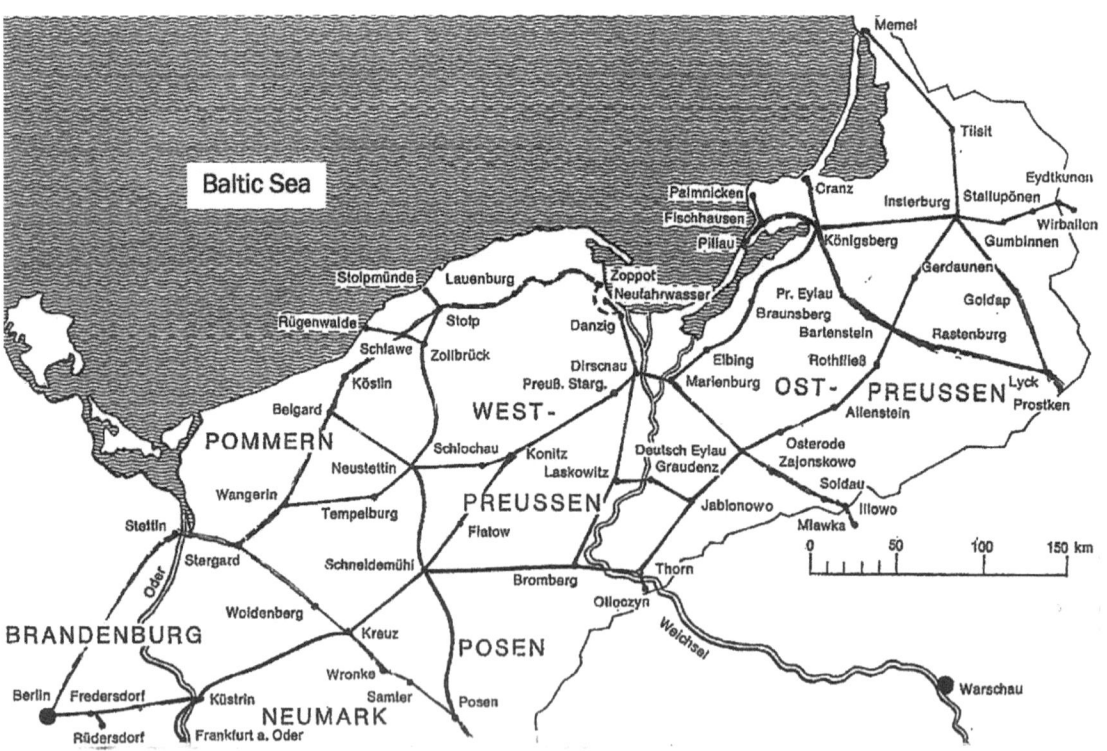

Eastern Railway Route Postmarks

20

The mail was transported in a closed compartment carriage attached to the normal train travelling from Berlin to Stettin to Bromberg. The postmark "Ostbahn" with the date and trip number was used from 1852 to 1853. It did not indicate the station where the letter had been collected.

The postmark "Ostbahn" was understood as a railway office postmark and there was no evidence of its having been used on a travelling post office. It was replaced by the more common line stamp postmark. The example below originated from a town far away from the Eastern Railway. Railway postmarks from other parts of the network are obviously not covered in my introduction.

Front of the cover

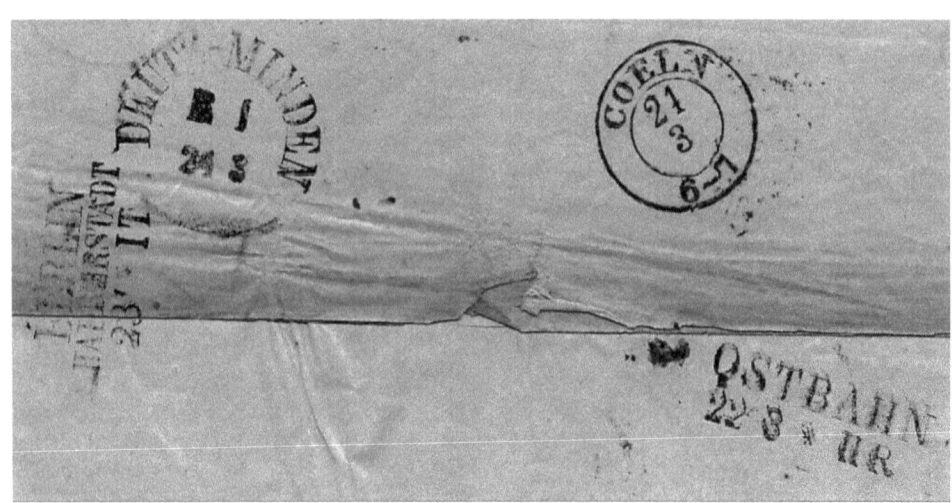

Railway postmarks on back of cover as per rail regulations

Until 1889, the transport of mail was conducted according to passenger train timetables. These timetables are now preserved digitally and available only through the Museum of Communications, Frankfurt, upon request.

A commercial letter dated 22 March 1851 received at the local office in Swinemünde, stamped 22 3 to Cöeln, with a local receipt postmark back stamped 24 3 in Cöeln. It was sent on the mail train from Bromberg via Stettin to Berlin. The Traffic Statement Postmark was "OSTBAHN / 22 3 *. II R" at the Stettin railway office.

Description of the "Traffic Statement Postmark" as per railway catalogue:

2.8.0.0   Special forms
2.9.0.0   Postmark of Route Statement
2.9.1.0   Office notes (Hint line) / Data line
2.9.2.0   Office notes / Data line / route address (examples not covered)

The hint line contained more detailed information on the designation of the postmark or the route.

The traffic information was, e.g., "EXTRAZUG" or railway names such as "OSTBAHN" or "Mountain Railway".

Single Line Railway Postmark type II BArGe reference 2.9.1.4

OSTBAHN/Eastern Railway: Trip *IIIR: Trip date 21 1 Postmark contains "traffic-technical statement", for example, "EXTRA TRAIN" or rail route name such as OSTBAHN. Postmark period 1851-1853

Commercial correspondence from Berlin to Wöhlen dated 25 January 1852. Received at Berlin city post office and delivered to Berlin Stettiner station (the starting point of the Ostbahn railway line). Sent on an outbound Ostbahn train returning to Bromberg and transferred to Halle outbound 25 1.

Back of the cover

Line Stamp Postmark type II BArGe reference 2.9.1.4

OSTBAHN/Eastern Railway; [Traffic Statement Postmark] Trip/Tour I R; Trip date 30 8 *; Postmark period 1851-1853

Stampless letter dated 29 August 1851 from Demmin, in Mecklenburg-Western Pomerania, addressed to Unter Gruslau Weißenfels (South Saxony-Anhalt). Mail was delivered to the nearest Post Forwarding Offices (Postspeditonsamt) III at the Eastern Railway Stettin station. Railway office postmark dated 30 8. The mail train travelled from Bromberg via Stettin to Berlin.

Reference line thought the series:

BArGe – Bundesarbeitsgemeinschaft Bahnpost

Line Stamp Postmark [Special forms] type II BArGe reference 2.9.1.4

OSTBAHN/111 Eastern Railway: Trip *IIR: Trip date 17 1. Postmark contains "traffic-technical statement". Rail route name OSTBAHN. Postmark period 1852-1853

c.1852 A letter received at Züllichau-Schwiebus, a Prussian province of Silesia, postmarked 16 4 *9-10. Mail collected at Kreuze station was delivered to Stettin station on the mail train returning to Berlin Stettiner station (Railway Office/Postspeditonsamt III).

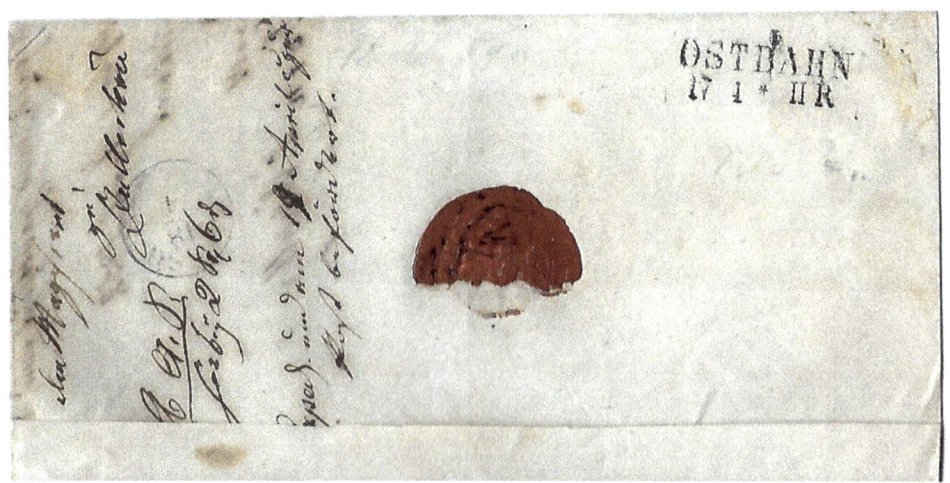

Traffic-technical statement (back of cover)
Route Network of the Eastern Railway Mail Train:
Bromberg-Posen-Stettin-Stettin-Berlin.

INBOUND FOREIGN MAIL ON THE EASTERN RAILWAY

Double Circle Postmark type II BArGe reference 2.1.3.0

BERLIN-HAMBURG; 29 5; Trip IR

Railway Station Mail Office Postmark

BROMBERG BAHNHOF; Date 30 5

Line Stamp Postmark type II BArGe reference 2.9.1.4

OSTBAHN/Eastern Railway; Trip number * IIIT; Trip date 29 5; Postmark period 1852-1853

Front of Cover

Commercial wrapper from Copenhagen, Denmark, received at the local office in Copenhagen 28 5 1852. Addressed to Inowrazlaw via Berlin Stettiner Station. Box Transit postmark applied top right. "Aus Denmark d. Mecklenburg" transferred 29 5 to Berlin-Hamburg IR. Mail transferred outbound (Berlin-Bromberg) on the Ostbahn Trip IIIT dated 29 5.

Receiving office at Bromberg Bahnhof stamped it 30 5, and it was delivered to Inowrazlaw by post rider.

The Prussian Railway Post Offices in Berlin

21

The Railway Offices (Postspeditonsamt) in Prussia, whose responsibility it was to transport letters and parcels up to 40 pounds by rail, started on 1 May 1849. In Berlin alone, there were four post offices, and this increased to over eight by 1858. They were responsible to the Berlin Court Post Office until 1 January 1851 when they were transferred to the newly created Head Post Office in Berlin. On 18 May 1862, the four Postspeditonsamt were finally integrated with the Berlin Postal Expeditions.

On 15 November 1850, the Prussian Post issued its first stamps. A four-ring stamp was issued for cancellation. The post offices were required to postmark stamps in the middle of the postage stamp. At the introduction, the numbers in a concentric ring were assigned alphabetically. A place stamp had to be stamped on the envelope to clearly identify the place of posting, as well as the posting date (later also the posting time). This would enable the mail transit time to be checked.

A special feature of the numeral postmarks assigned to the post-forwarding agencies was that they were not assigned to one location but to the post-forwarding office responsible for the transport of letters and parcels in that part of the rail network. A post forwarding office was located at a larger train station, and only a few are covered by my Eastern Railway series. The following is a list of post forwarding offices opened since 1850; they gradually increased as the service became ever more popular with the expanding population across Prussia.

The railway stations that had been directly connected to Berlin by rail since 1850 used the ring postmark. Travelling postal officers worked on a number of direct routes. The responsible railway post offices in Berlin used their own number postmarks below to cancel the postage stamps.

THE PRUSSIAN RAILWAY POST OFFICES IN BERLIN

The following chapters cover only the following ring cancels.

BERLIN 3		RING CANCEL 103 (C. 30 TYPES)
Post Forwarding Office I	Berlin – Potsdam train station	Ring Cancel 104 (c.8 types)
Post Forwarding Office II	Berlin – Anhalter train station	Ring Cancel 105 (4 types)
Post Forwarding Office III	Berlin – Stettiner train station	Ring Cancel 106 (3 types)
Post Forwarding Office IV	Berlin – Silesian train station	Ring Cancel 107 (c.8 types)
Post Forwarding Office XI	Bromberg/ Dirchau/ Königsburg	Ring Cancel 1768 (c.4 types)
Post Forwarding Office XII	Marienburg/ Königsburg/ Dirchau	Ring Cancel 1781 (c.4 types)

The official names Post-Speditionsämter and Post-Speditions-Bureaus were renamed in the Official Gazette Order 20 of 29 January 1856 as railway post offices. The practice of applying ring cancellations to postage stamps was discontinued from 1 April 1859.

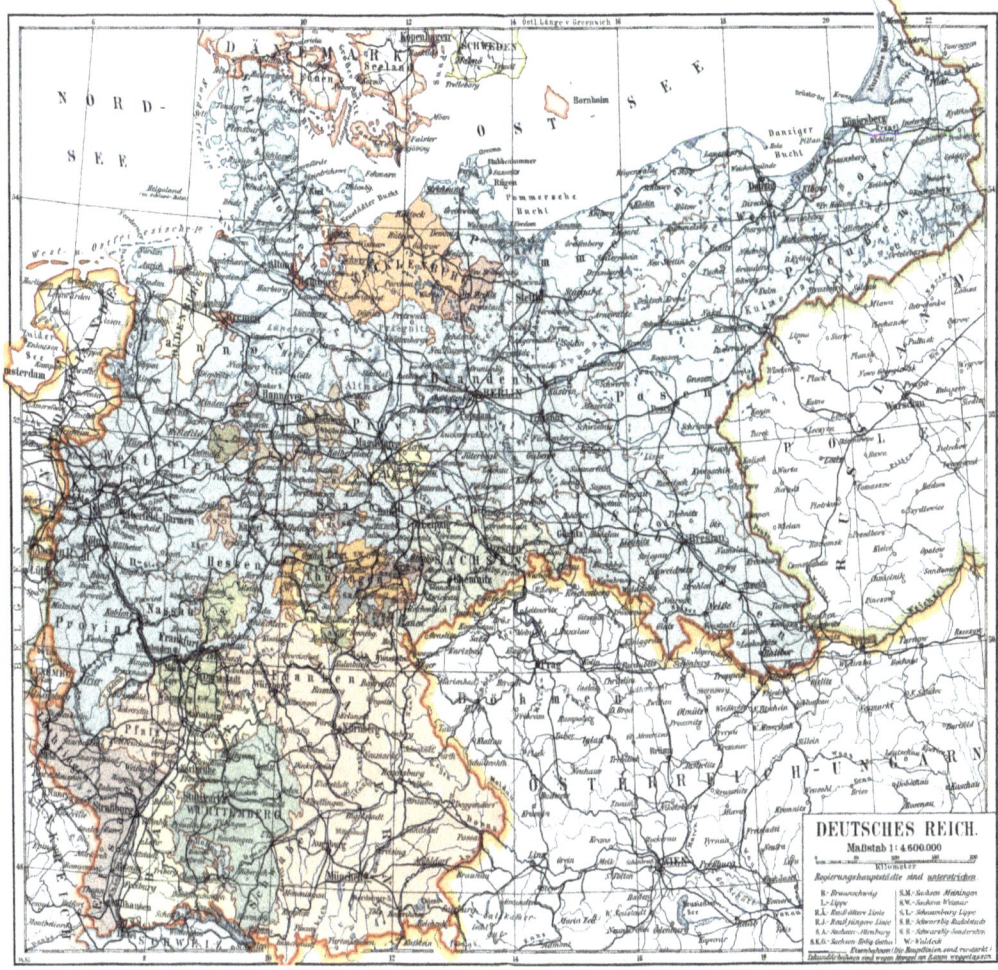

The density of the railway network from Berlin 1893–1918

Berlin Main Stations – Earliest Concentric Postmarks

22

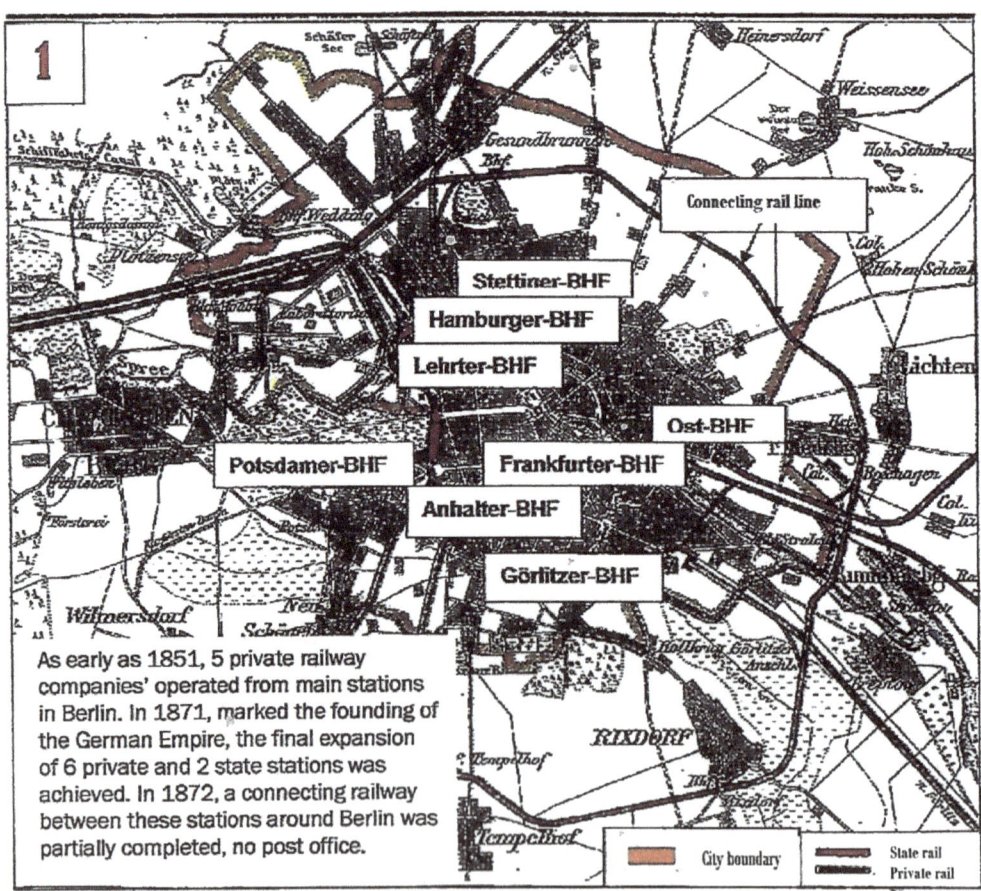

It is noteworthy that the cancellation of the postage stamps with the respective numeral postmarks often took place on the train; in some instances, letters and parcels received a further single-line stamp indicating the place of posting, such as "Stettin". Several stamps with the same number were used simultaneously on the individual lines of a railway post office. This helps to explain the relatively high number of different types of these number stamps. Occasionally the cancellation by the number stamp was not done in black, as prescribed, but in red. Red stamping ink was mainly used to mark parcel deliveries, and it could happen that the postal worker used the wrong stamp pad to stamp the mail.

Concentric Postmark 103

Berlin Railway Office (c.30 types); Dispatch date 19 11* 1-2Nm

Pre-paid letter from Berlin to Basel. Berlin City Post Office XII on 19 November 1858; delivered to the Berlin Railway station, "Berlin Railway Office 103" for dispatch. Prussian Foreign Letter rate (Franco – post paid) postage stamps 3 Sgr. Catalogue reference (MiNr.6a)

Concentric Postmark 103

Berlin Stettiner Station Post Office

MiNr. 2

MiNr.4a

MiNr.8b

MiNr.5b

58 HISTORY OF THE EASTERN RAILWAYS CONSTRUCTION AND EXPANSION

Concentric Postmark 104

Berlin Post-Sped.-Amt Nr.1

(Forwarding Office Berlin Silesian train station) (c. 8 types)

c.1856 Pre-paid letter posted at "Berlin Potsdamer Bahn": 18 2 * 4-5 and delivered to Berlin Silesian station; railway office postmark 104; addressed to Breslau. Prussian Inland Letter rate post paid with MiNr.2b on Michel stationery catalogue U 12

Concentric Postmark 104

Berlin Silesian Station Post Office 1

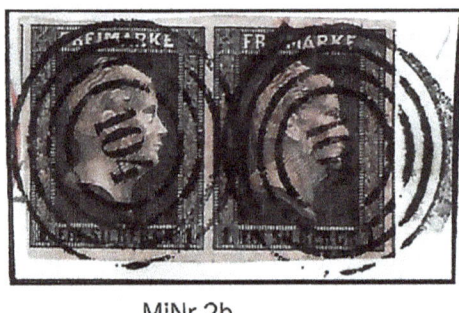  

MiNr.2b   MiNr.3   MiNr.8ax

The railway network for Berlin Post-Sped.-Amt Nr.1 1851-1860

The railway network for Berlin Post-Sped.-Amt Nr.II 1851-1860

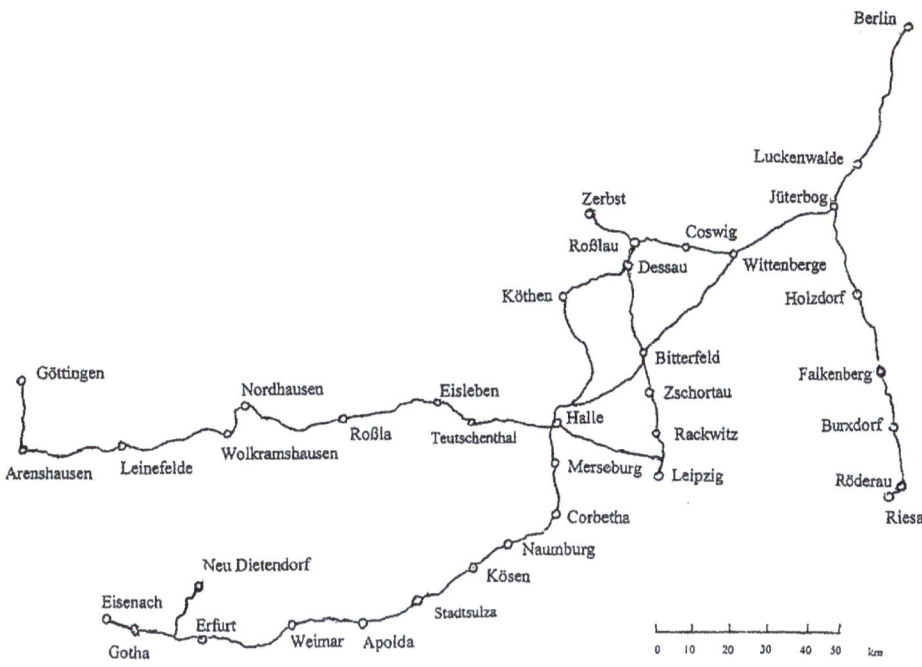

Concentric Postmark 105

Berlin Post-Sped.-Amt Nr. II (Forwarding Office Anhalter train station) (c.4 types)

A Prussian letter received at Berlin Anhalter rail station was postmarked 10 3 * 6-7 and transferred to a travelling post office. It was postmarked with a ring cancellation 104.

Signed
Kaustaun BPP.
Back of the cover

c.1890 Berlin Anhalter railway station

Concentric Postmark 106

Berlin Post-Sped.-Amt Nr.III (Forwarding Office Berliner Stettiner Station) (3 types)

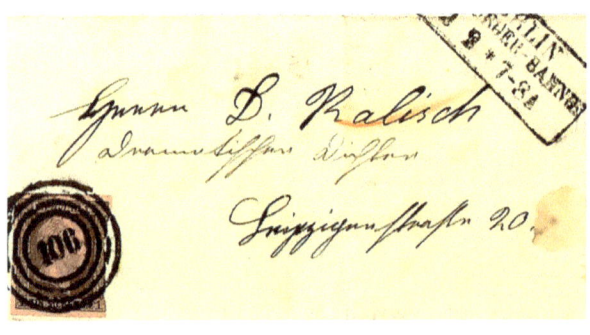

A letter from Berlin received at the Berlin Hamburger-Bahn station Office 2 2 *7-8 A. Delivered to the Berlin Stettiner Railway station. Postmarked at the Berlin Railway Office for dispatch. Prussian Inland Letter rate 1 Sgr. affixed; Michel Deutschland catalogue ref. MiNr.2b

Concentric Postmark 105

Berlin Post-Sped.-Amt Nr. II (Forwarding Office Berlin Anhalter station Post Office) (4 types)

MiNr.1   MiNr.3   MiNr.4a   MiNr.5a

Concentric Postmark 107

Berlin Post-Sped.-Amt Nr. IV (Forwarding Office Berlin Silesian train station) (c.8 types)

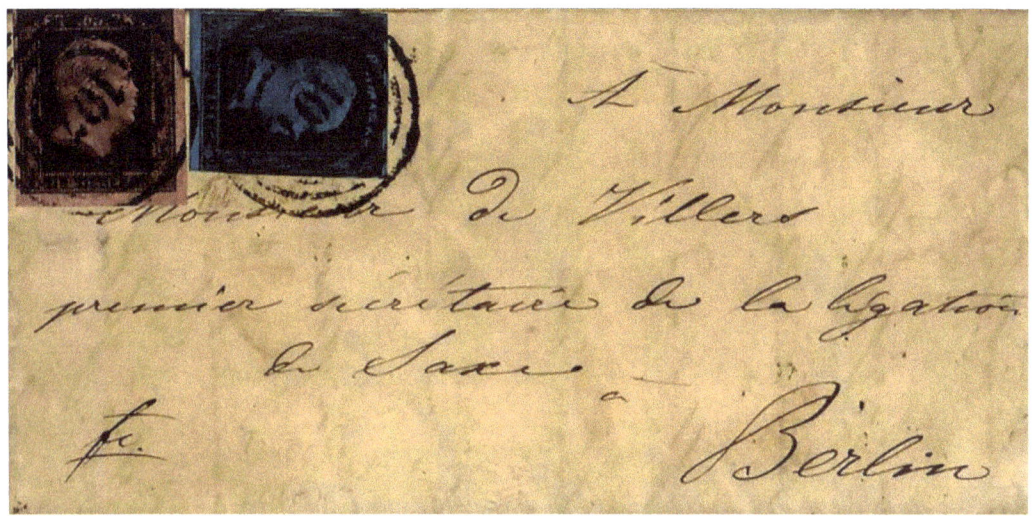

c.1852 A letter addressed to a recipient at the Saxon embassy in Berlin. Delivered to the Berlin Silesian Railway office for dispatch, postmarked with concentric 107. Prussian Inland Letter rate 3 Sgr; postage catalogue reference MiNr.2+3

Concentric Postmark 107

Berlin Post-Sped.-Amt Nr. IV:

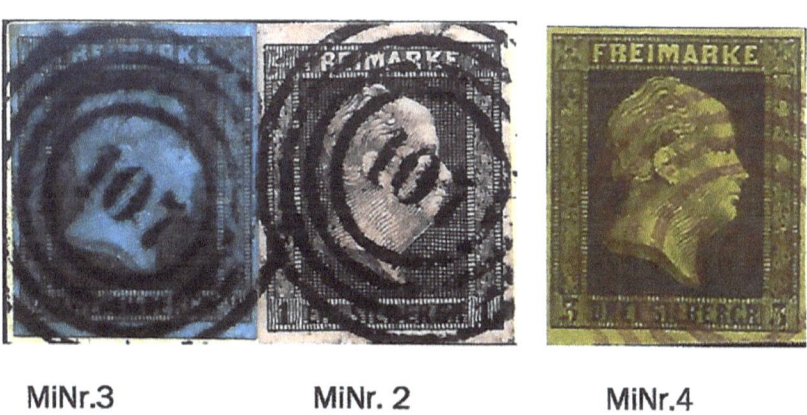

MiNr.3   MiNr. 2   MiNr.4

As the terminus of both the Silesian and the Eastern Railway line, Schlesischer Bahnhof soon developed into Berlin's "Gate to the East". Until World War I, trains ran from the German capital via Königsberg to St.Petersburg (Nord Express) and Moscow, as well as Vienna, Budapest and Constantinople via Breslau and Kattowitz. During the Anti-Jewish pogroms in the Russian Empire, numerous Jewish refugees arrived here en route to the emigration harbours in Hamburg and Bremerhaven to board ships to the United States.

Imperial Railway Post Office Berlin receiving Foreign Mail 23

Official Mail postcard originating from Peking, postmarked 17 4 13

Official post postcard from Hankau addressed to the Kaiserliche Bahnpostamt 18, Berlin 17, and postmarked 3 5 12

Berlin Stettiner Railway Station for the Postspeditonsamt III (OSTBAHN Routes)

24

The Eastern Railway and its network of stations were connected to the Berlin-Stettiner railway station. The station was built by a private company and formally opened on 16 August 1843. The station name appeared on "Ring" and "Frame Cancel" postmarks during the early years. On 15 November 1850, the Prussian postmark numeral 106, which belonged to Postspeditonsamt III, was introduced for the Stettiner and Hamburger stations as well. The concentric postmark was used for a short time at both stations, from 8 December 1850 to 18 June 1851.

Early period station postmark

Rebuilt 1902 facade of Stettiner Railway Station Berlin Postspeditonsamt III

Frame Cancel Postmark: Railway Office

ANHALTHER BAHNH. BAHNPOST N°2; Dispatch Date 21 7.81

The Anhalter (Bahnhof) was what we know as the Railway Post Office 2 in Berlin. Collectors do not know whether the postmark below was stamped at the station office or was an experimental postmark as it is designated as "Rare". The station was responsible for the routes to Coethen-Leipzig, Wittenberg-Leipzig and Roederau.

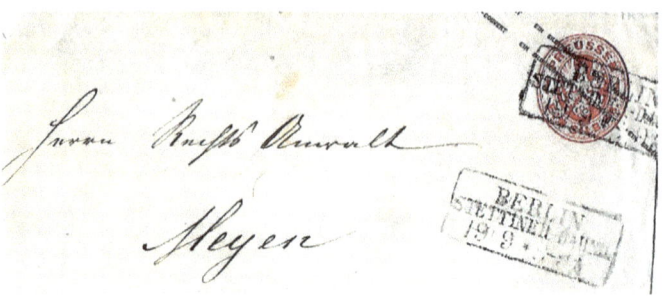

FRAME CANCEL POSTMARK: Prussian Stettiner Railway station postmark

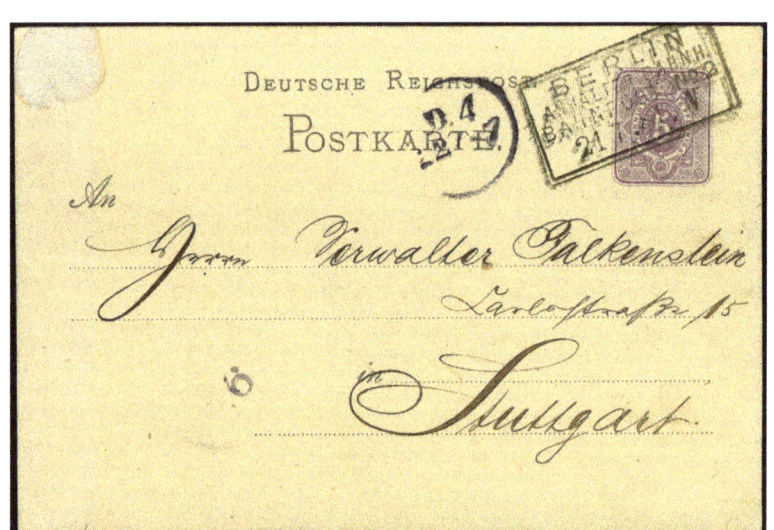

"Very Rare" Pre-paid postcard dated 21 July 1889 from Berlin addressed to Stuttgart. Delivery postman's thimble postmark 22 7 stamped on the front.

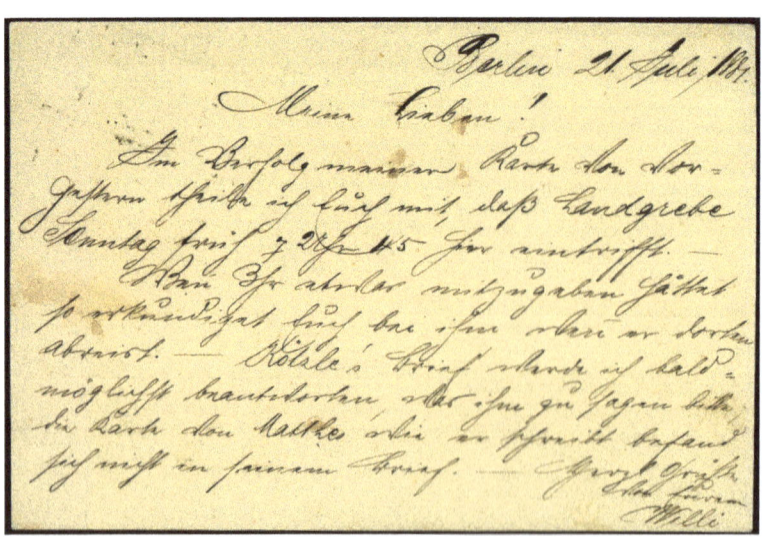

Back of the postcard

BERLIN STETTINER RAILWAY STATION FOR THE POSTSPEDITONSAMT III (OSTBAHN ROUTES)

Frame Cancel Postmark: Railway Office

BERLIN STETTINER-BAHNH: Dispatch date 3 2 * 5-6; width 35.00 mm; Postmark period 1865

Prussian pre-paid letter from Berlin addressed to Danzig. Mail delivered to Stettiner railway station, postmarked at the railway office and dispatched to Danzig.

Stettiner Bahnhof postcard from Berlin 30.4.1901 addressed to Knittelfeld

THE RAILWAY POST OFFICE III STETTIN

In 1842, the Stettiner Bahnhof opened as the terminus of the railway line to the then German city of Stettin. It connected Berlin with the Stettin seaport, which was one of the headquarters of forwarding office No. 3 at Stettin Station. The concentric numeral postmark 106 was used at the station office as a cancellation stamp until the end of January 1859

Stettin had at least two post offices, the oldest of which had existed since 1682. Stettin had used a double circle postmark "BERLIN STETTIN" since 1845. In addition, the postal authorities used a three-line frame cancel postmark with a few variations (example below).

Line Stamp Railway Postmark type II BArGe reference 2.1.3.0

STETTIN - BERLIN; Tour/Trip II; Trip date 5 3; Postmark period 1851-1904

(Front Cover) c.1857 Express Letter from Stettin addressed to Berlin. Mail received at the Stettin rail office for Railway Post dispatch office Nr. III (Concentric 106). Berlin postmarked Trip II. Receipt postmark back stamped Berlin 5 3.

(Back Cover)

Departed Stettin Station at 12.35 a.m. → Passow - Angermünde - Neustadt - Berlin. Local delivery office double ring postmark back stamp 5 3, time 4.5 Nm. (before noon).

Route Postmark Catalogue for Berlin – Stettin

ROUTE	REFERENCE	PERIOD	NOTES
BERLIN – STETTIN	2.1.3.0	1852-1887	Diverse
BERLIN – STETTIN	2.1.3.5	1874-1886	
BERLIN – STETTIN	6.1.1.0	1851	
BERLIN – STETTIN	7.1.1.0	1888-1920	Diverse/First distribution date 1.5.1898
BERLIN – STETTIN	7.1.2.0	1890-1918	First distribution date 13.7.1887

Dear Readers and Researchers

In 1973 collectors and researchers joined together to form a society and collect historical data in the field of Railway Post Offices in the Federal Republic of Germany. This was the foundation of the RPO study group. The society publishes a journal twice annually. Those who have joined are from other European countries and some even from overseas. Anyone who is interested in RPOs can apply for membership of the German RPO Society. http://www.uqp.de/bahnpost/englisch/starte.htm

Stettin Railway Station Office Postmark

25

When mail volume increased, the Station Post Office (Bahnhofpostamt) also accepted and postmarked mail of all kinds. The postmark with the station name in a cancel frame was applied, and sorted mail was dispatched to its next destination. An administrative change to the station postmarks can be traced with the wording changed to Stettin Bahnhof in 1852.

In 1860, "Station expedition" (Bahnhofsexpedition) and "Speditions-Comptoir" in Stettin were changed to station post office (Bahnhofpostamt). From then on, "Stettin Bahnhof" stamps were used. This new rectangular stamp was available in several types, recognisable by the inclusion of the word "station".

Frame Cancel Postmark

STETTIN BAHNHOF: Dispatch date 23 5; Postmark width 33.50 mm: "BAHNHOF" width 20 mm; Postmark period 17.11.1853 to 26.11.1873

A total of three types of the numeral postmark '1439' (Stettin) can be verified.

c.1857 Commercial correspondence addressed to Berlin received a ring postmark at Stettin City Office, "1439" concentric cancel and a registration stamp at the receiving office. Letter was delivered to Stettin Railway station postmarked "Stettin Bahnhof"; mail office postmarked 23 5 (1858). Prussian Inland Letter rate for registered mail; postage 2 Silbergroschen postage paid.

Until 1850, the postal system in Prussia was organised centrally. All post offices were answerable directly to the General Post Office in Berlin. On 1 January 1850, the administration was decentralised, and 26 senior post offices were created for the administrative districts. One of them was Stettin. It was the task of a head post office to relieve the pressure on the general post office in Berlin.

As the new intermediate authority, they had to take care of the administration of the postal service's working in their district, the personnel management and the treasury. The Chief Postal Director ran the administration in the district entrusted to him independently and on his own authority.

The Prussian post office class designation that had existed since 1852 was dropped with a ruling in the official gazette of 25 July 1924.

This post office building, completed in January 1904, was the seat of not only the post office but also the telegraph office and the post office in Stettin 4. It was built under the supervision of Postal Construction Council director Hintze in a neo-Gothic style.

Imperial Post Office Stettin. The city of Stettin, Pomerania. The Kaiserliche Head Post Office. Postcard dated 5 January 1904 addressed to Schwedt i.Pom.

Line Stamp Railway Postmarks of Stettin and Postmark Catalogue

26

2.1.0.0 THREE LINE POSTMARKS (SCRIPT: ROMAN TYPE WITH SAN-SERIF)

    2.1.1.0 Departing/ Destination / Data line
    2.1.2.0 Left empty / Data line/ Destination
    2.1.3.0 Departing/ Data line / Destination (the example below)
    2.1.4.0 Departing/ Data line / Section
    2.1.5.0 Task place/ Data line / route
    2.1.6.0 Route/ Data line / Place of posting
    2.1.7.0 Route/ Data line / Departing Office
    2.1.8.0 Departing Office / Data line / route

2.2.0.0 THREE LINE POSTMARKS, FONT: (SCRIPT: FINE ROMAN WITHOUT SANS-SERIF FONT)

    2.2.1.0 to 2.2.1.0:
    2.1.1.0 to 2.1.8.0, respectively

Line Stamp Postmark type II BArGe reference 2.1.3.0

STETTIN - BERLIN; Tour/Trip II; Trip date 13 9; Script common Roman; Departing station / Data line / Arrival station, Postmark period 1851-1904

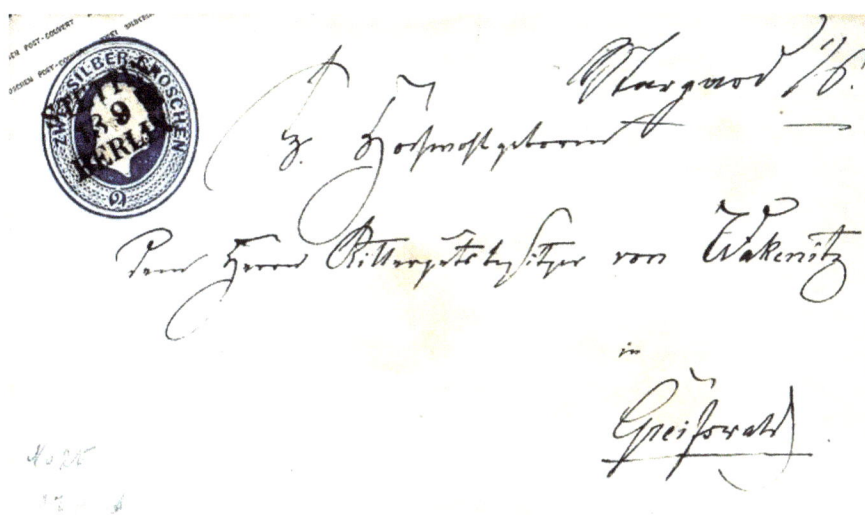

c.1855-57 Pre-paid Prussian letter from Stargard. Addressed to Greifswald 16, delivered to Stettin station mail train trip II outbound from Stettin to Berlin.

Prussian Inland Letter rate; 2 Sgr. postage paid on Michel Ganzsachen U 12.

Frame Cancel Postmark: STETTIN

The station stamp was used to validate the mail posted directly in the train letterbox. This saved the handwritten entry of the place of posting. The station postmark appeared first in black, more often in blue.

Line Stamp Postmark type II BArGe reference 2.1.3.0

BERLIN - CÖSLIN; Tour/Trip IT; Trip date 28 2; Script common Roman; Departing station / Data line / Arrival station, Postmark period 1858-1904

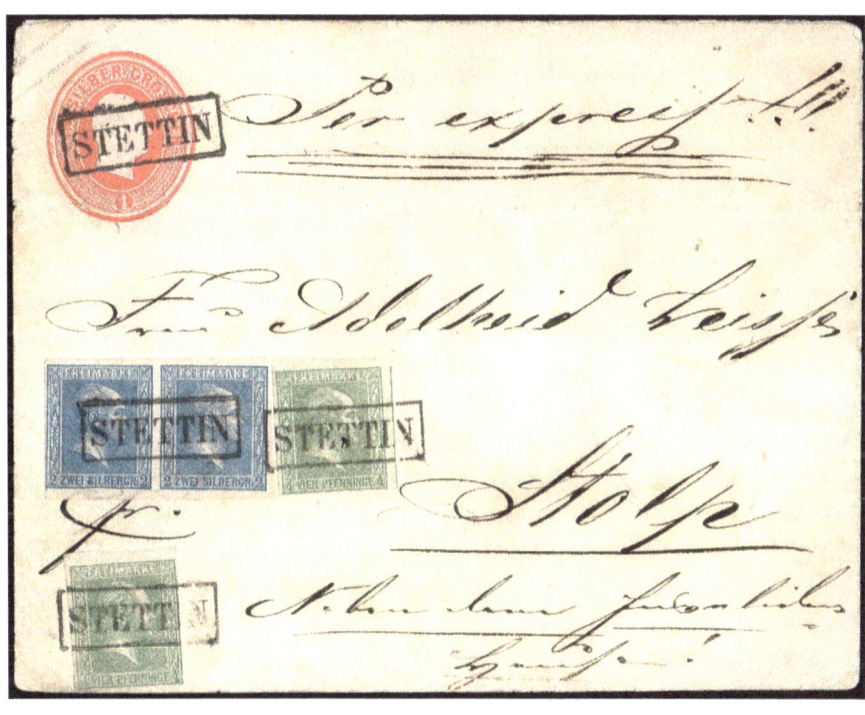

Express pre-Paid letter from Stettin to Stolp. Letter received at Stettin office, Single Line cancellation. Mail delivered to Stettin station and postmarked on a TPO travelling outbound from Berlin via Stettin to Cöslin on Michel Ganzsachen U 11.

Back stamp supporting railway postmarks

Line Stamp Postmark type II BArGe reference 2.2.3.

STETTIN - BŰTZOW; Tour/Trip T; Trip date 31 12; Script common Roman; Departing station / Data line / Arrival station, Postmark period 1872-1882

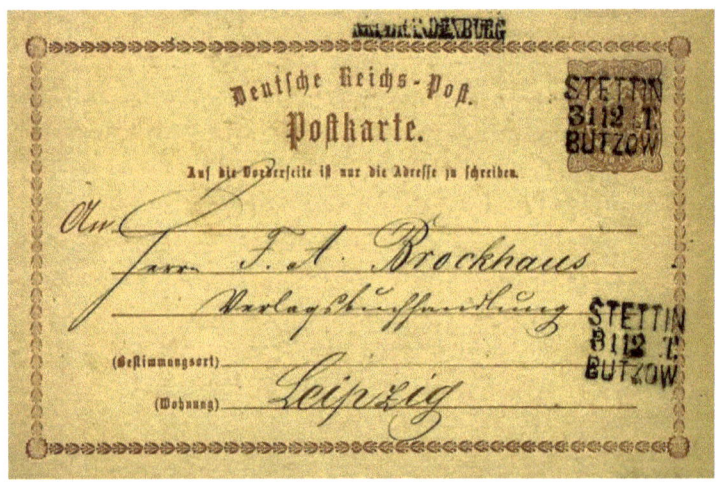

Private postcard from Neubrandenburg addressed to Leipzig. Mail received at the Neu Brandenburg station was delivered to a mail train outbound 1T from Stettin to Bützow.

c.1890 Gruss aus Pasewalk station. Postcard dated 12.12.97 addressed to Magdeburg

Railway Office: Station (City) Expedition

27

The following example is the earliest known railway station postmark close to the Baltic Sea port. It was in use long before the Eastern Railway was politically conceived.

It highlights the origins of the railway office postal service. The first railway connection was completed between Stettin and Berlin on 15 August 1843. The travelling time was estimated at 4½ hours, and two trains a day ran in either direction. The postal authorities in Stettin made use of the faster means of transport by rail and set up a post office in the station.

1844. ESTABLISHMENT OF A STATION EXPEDITION

20/12/1844. Very early use of this new line stamp postmark. It consisted of three lines: station name/office/date. It was postmarked at the "Stettin Railway Expedition" office. Usually, this postmark is either poorly chipped or dirty.

Railway Station Postmark – SPED.COMTOIR No.3 STETTIN 28

Stettin had at least two post offices, the oldest of which had existed since 1682. Stettin had used a double circle postmark since 1845, BERLIN STETTIN. The postal authorities changed it to a three-line frame cancel postmark with a few variations.

One of the post offices was a Bahnpost-Expedition office (established in 1843). Here a three-line frame postmark STETTIN BAHNHOF was used as the issue stamp. There are several types of this.

Ten years later, in 1854, the Stettiner Bahnhof-Expedition was renamed "Speditions Comtoir N°3" and functioned in addition to the Eisenbahnpostamt 3, which had been set up in Berlin in 1849 and managed the railway lines to the north.

A new postmark "Speditions - Comtoir" was issued for this office in 1854.

There was also a Post Speditionsbüro (or Comtoir) of the Post Speditionsämter Counter No. 3 in Stettin. This used the numeral postmark 106 to stamp the postage on mail items and a three-line frame stamp SPED.COMTOIR No.3 STETTIN (Comtoir/Office counter) (see Figures 23 and 24).

Frame Cancel Postmark

SPED.COMTOIR N°3 STETTIN; Trip date * 6 7, Width 42mm x 15mm; Postmark period: 01.06.1857 to 30.07.1857

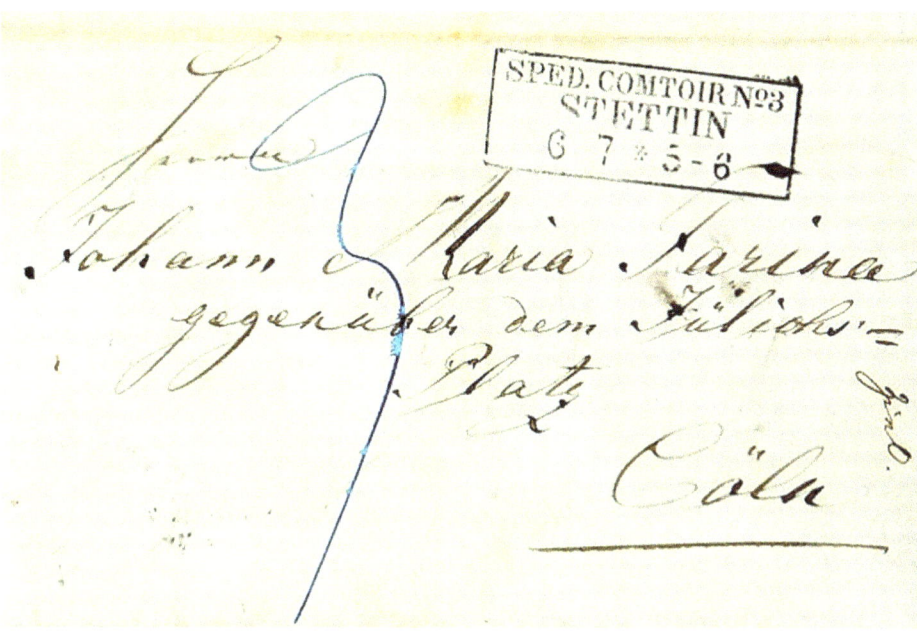

c.1857 Seldom seen postmark. Letter from Stettin addressed to Cöln. Mail train Stettin to Berlin and Deutz-Minden to Cöln. Prussian fee collection written in blue.

Frame Cancel Postmark

SPED.COMTOIR N°3 STETTIN; Trip date 30 11 * 6 7: Width 42mm x 15mm; Postmark period: 01.06.1857 to 30.07.

Letter from Stettin addressed to Warschau. Mail delivered by a conductor mail train from Stettin to Warschau; receipt office front stamped in red 11/30 and a thimble postmark by postman dated 30/11 Warschau.

Prussian Foreign Letter rate paid 3 Sgr. Michel Deutschland catalogue reference MiNr.12a

Sender's detail and delivery office postmark back stamped.

NORTH GERMAN CONFEDERATION POST CORRESPONDENCE

In 1841 the Breslau-Schweidnitz-Freiburger Railway Company (BSF) was founded in Breslau. It developed into an important transport company, with a rail network of over 600 km in length until its nationalisation in 1886. On 15 May 1877, the port city and Pomeranian capital, Stettin, was connected.

The Eisenbahn Post-BŰR route (railway post office) Nr.3 network runs from Breslau in a south-north direction and connects the two cities of Breslau and Stettin. After the establishment of the Deutsche Reichsbahn, the route network was assigned to the Reichsbahn Management, Breslau, including Küstrin and Stettin stations.

Frame Cancel Postmark

STETTIN EIS.POST-BŰR. No. XIV; Trip date 17 3 74

Post-paid printed matter rate ½ Groschen. Correspondence card from Stettin addressed to Dresden.

The Railway network for the Stettin Railway Office XIV 1871-1920

Eisenbahn Post Büreau 3 Stettin (Deutsche Reich)

29

Frame Cancel Postmark

EISB.POST-BŰR. STETTIN 3; Trip Date 11 1 * 3 4 (afternoon)

Commercial invoice post paid inland printed matter rate ½ Groschen. Stettin Network station.

On letter piece ½ Groschen and 2 Groschen: EISB.POST-BŰR.3 STETTIN 2 4

Other Connections of Stettin Station

The Railway network for the Breslau Railway Office III 1871-1920:

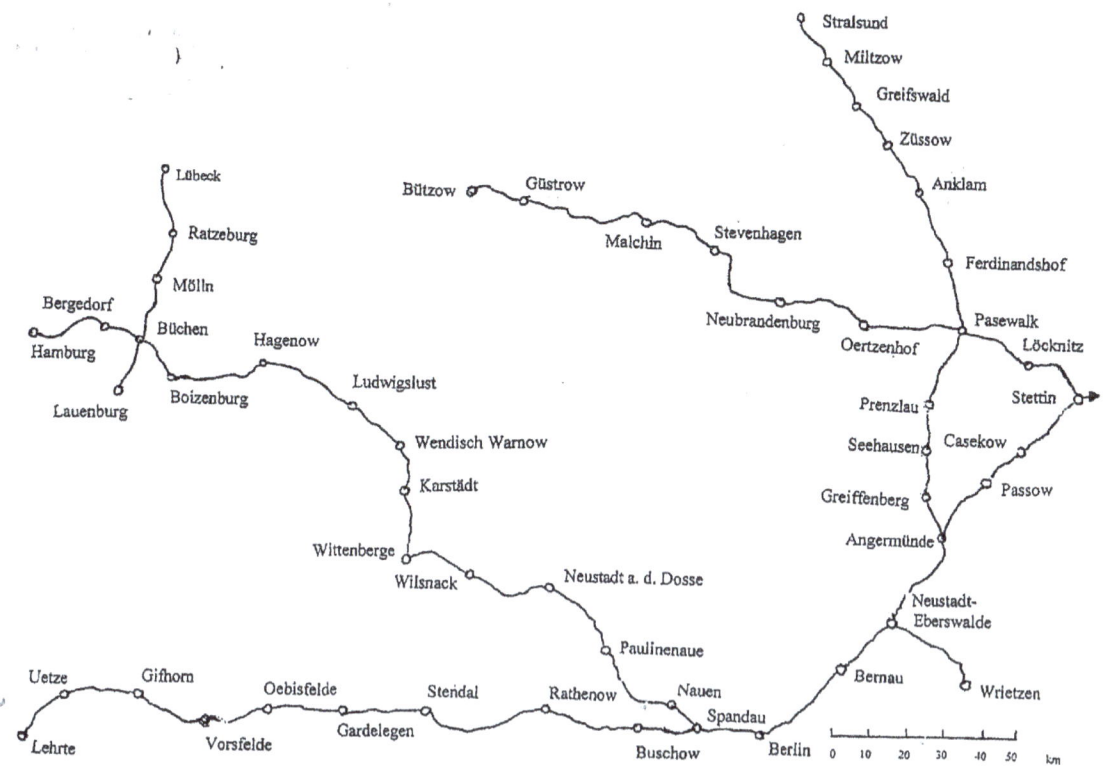

The route network for the Eisenbahn Post-BÜR (railway post office) Stettin. Nr.3. Post traffic was quite extensive between the two stations of Breslau and Stettin.

Map of Pomerania

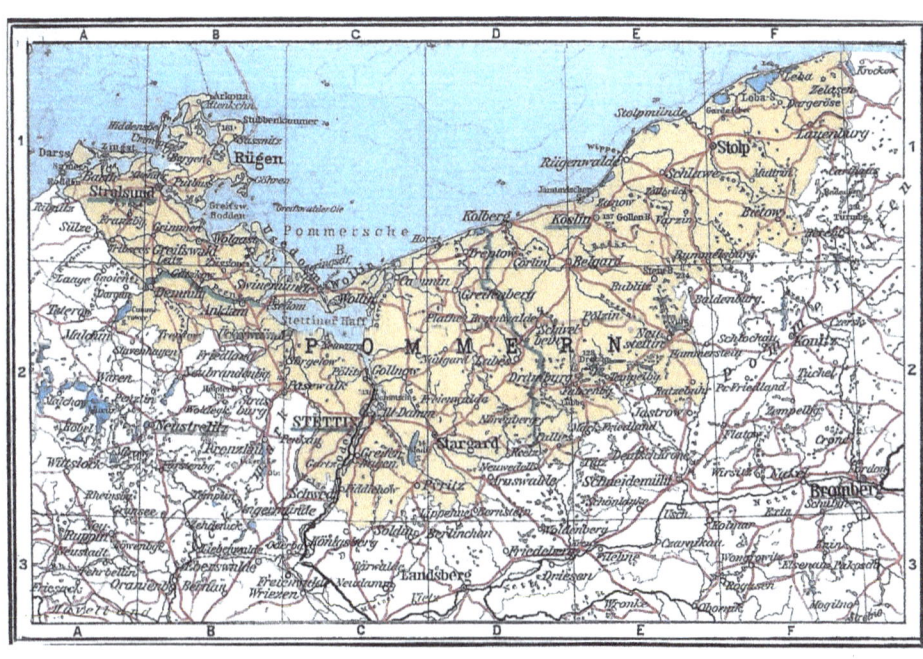

Railway lines connected Stettin to all parts of the Pomeranian coastline and the important stations of Stargard and Bromberg.

Small Oval Postmarks of Stettin 1883-1921

Route statement against Oval I to Oval V always in curved format, with Oval VI double-spaced parallel; Wording of "BAHNPOST" horizontally below the route statement.

7.1.0.0 Oval Figures 1

Data lines: ZUG, Train number / Day, Month, Year

7.1.1.0 ZUG (Roman)
7.1.2.0 ZUG (sans serif fonts)
7.1.3.0 Z instead of ZUG

DEUTSCHER REICH – STETTIN

Small Oval Postmark type II BArGe reference 7.1.1.0

STETTIN - KOLBERG: Train Number ZUG 849; Trip date 19.6.17; Postmark period 1905-1921

STETTIN – KOLBERG RAILWAY

The route consisted of two historical sections. The older one from Belgard to Kolberg was opened to traffic on 1 June 1859 by the Berlin-Stettiner Railway Company, together with the Stargard-Köslin railway. The second section between Neustettin and Belgard had been in service since 15 November 1878, managed by the Prussian State Railway as part of the Eastern Railway network. From 1920, the Deutsche Reichsbahn took over as the successor to the Prussian State Railways.

Small Oval Postmark type 7.I BArGe reference 7.1.2.0

BŰTZOW - STETTIN; Train number ZUG 5; Trip date 12 12.97; Postmark period 1891-1914

Postcard dated 12.12.97 addressed to Magdeburg, with a receipt postmark back stamped in Magdeburg 12.12.97. Stationary delivered to Bützow station on arrival in Stettin.

c.1900 Berlin Stettiner station. Unused postcard

Large Oval Postmark type 7.II BArGe reference 7.2.4.6

STETTIN - KOLBERG: Train number ZUG 844; Trip date 31.8.28; Postmark period 1928-1943 VZ 4/Diverse/First distribution date 20.7.1928

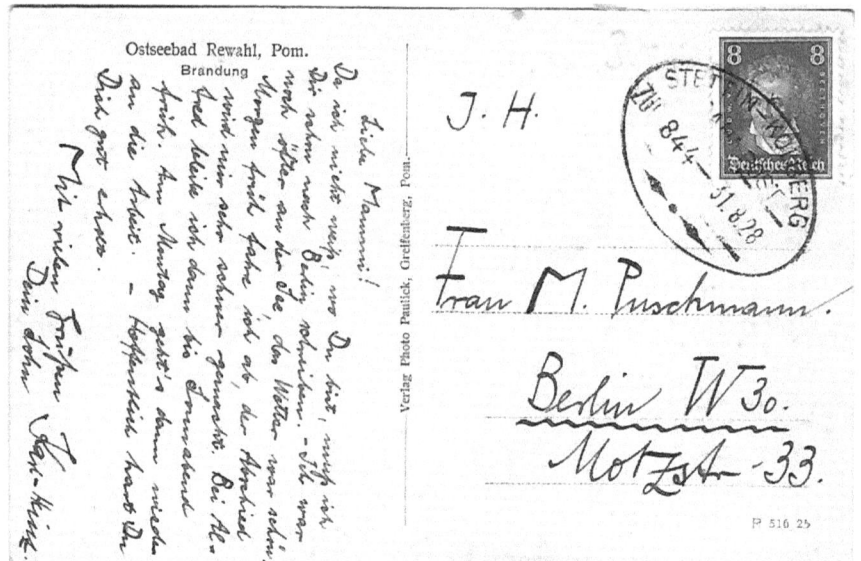

Stationary card from a sea side town of Rewahl, Pomerania addressed to Berlin. Departed Kolberg 8:10 arrival Stettin 12:11 mail transfer to a Berlin bound train.

Postcard from the seaside town of Rewahl, Pomerania, addressed to Berlin. Departed Kolberg 8:10 arrival Stettin 12:11. Mail transferred to a Berlin-bound train.

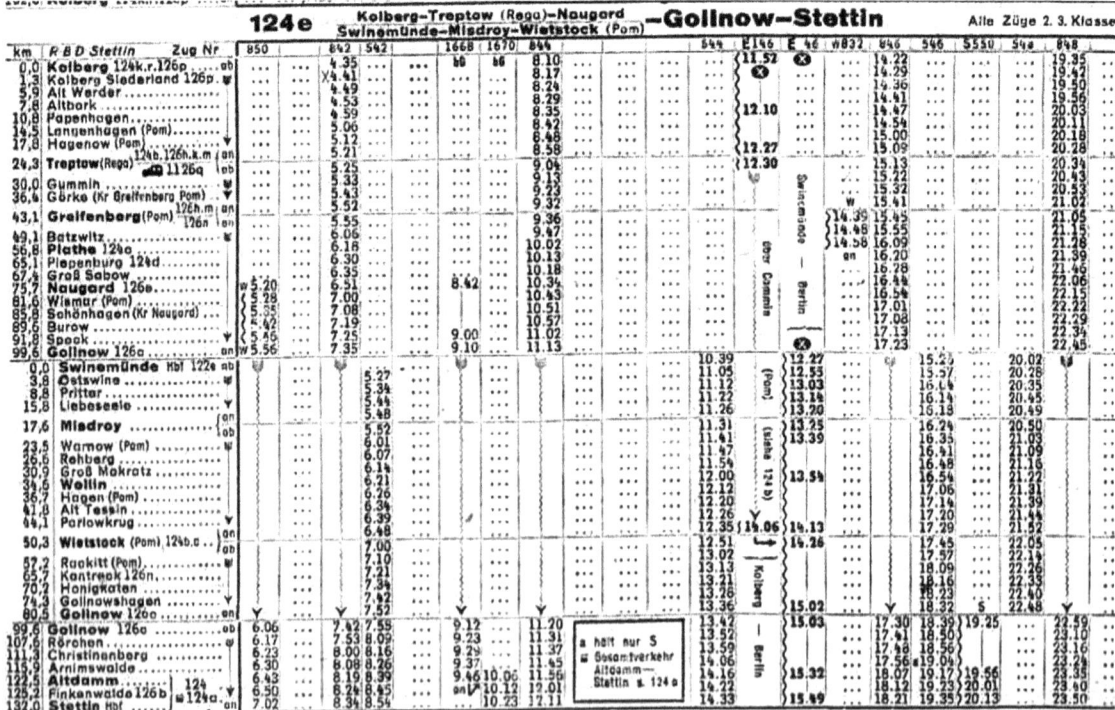

SMALL OVAL POSTMARKS OF STETTIN 1883-1921   83

Large Oval Postmark type 7.II BArGe reference 7.2.1.0

STETTIN - Großziegenort; Train number Z. 498; Trip date -1.5.16; Postmark period 1910-1930

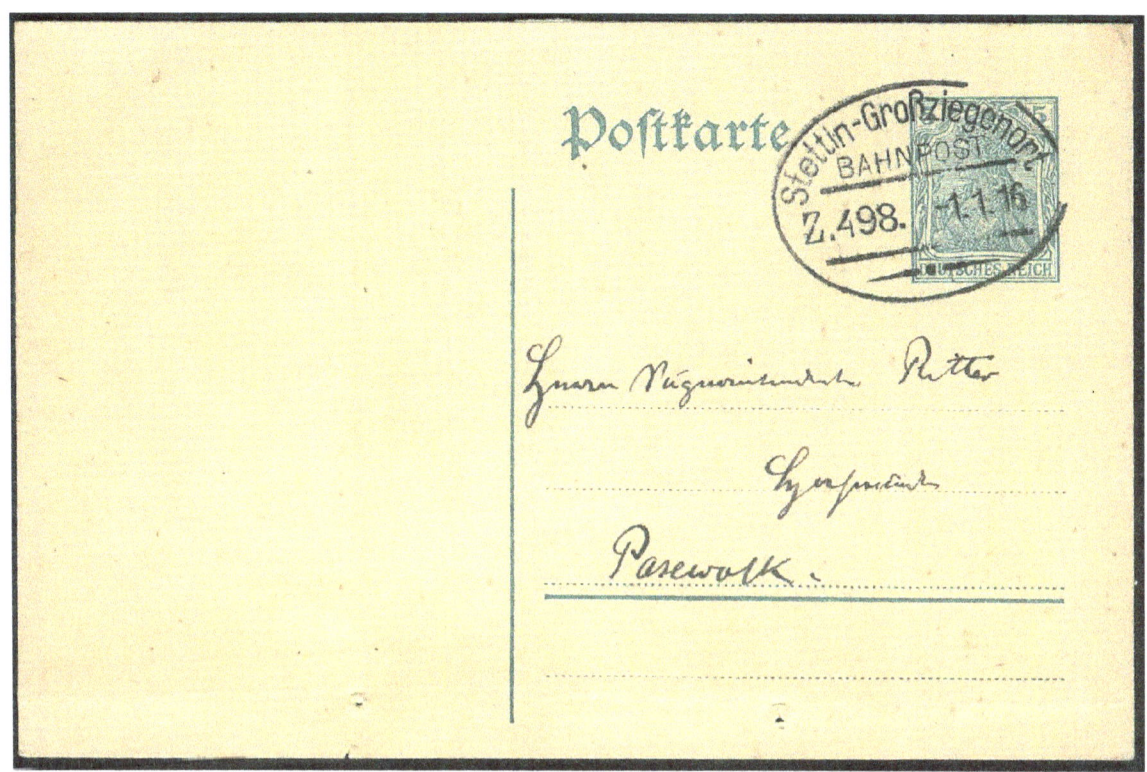

Pre-paid postcard addressed to Pasewalk. Mail train outbound from Stettin Main Station → Stettin–Torney → Züllchow – Frauendorf – Gatzlow – Stolzenhagen-Kruizwieck → Odermünde → Messenthin – Pölitz – Jasenitz – Damuster – Königsfelde-Wilheimsdorf → Ziegenort station.

Map showing branch railway route Stettin – Großziegenort:

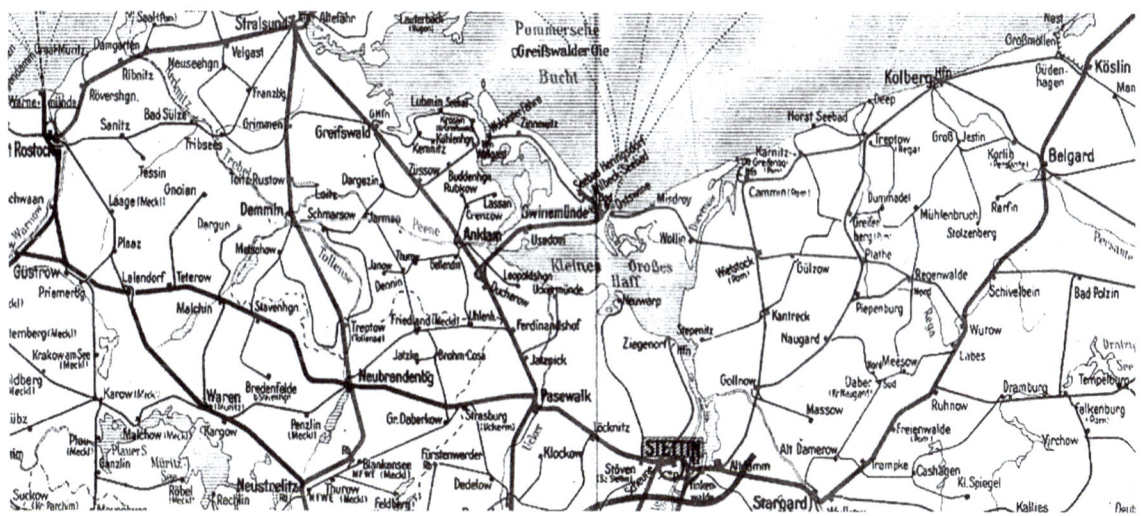

84 HISTORY OF THE EASTERN RAILWAYS CONSTRUCTION AND EXPANSION

Large Oval Postmark type 7.III BArGe reference 7.3.4.1 index 'a'

STETTIN - KOLBERG (OSTSEEBAD); Train number Z. 848; Trip date 26.6.39; Postmark period 1937-1944

Mail train returning from Kolberg main station at 19.35 arrived in Stettin at 23.50.

c.1920 Kolberg main station. Unused postcard.

Large Oval Postmark type 7.II BArGe reference 7.2.1.0

STETTIN – SWINEMÜNDE: Train number Z. 54; Trip date 29.8.23; Postmark period 1910-1944

West Pomeranian district to Breslau. Return mail train departing Swinemünde via Pasewalk and Ducherow to Stettin.

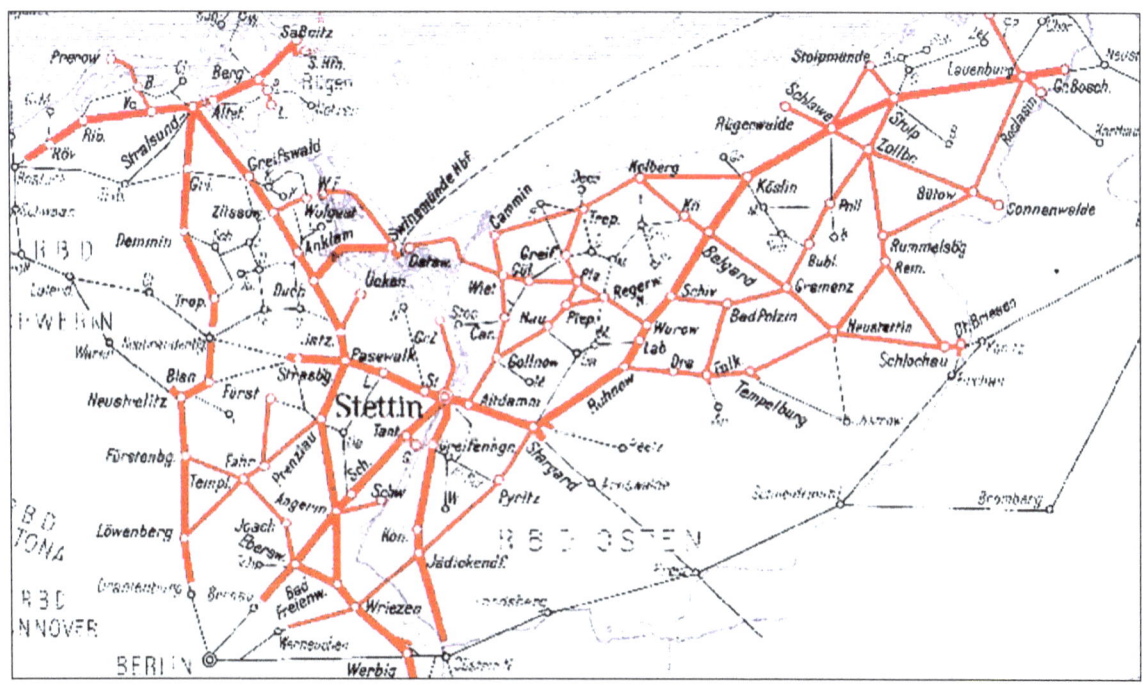

The Route Network of the Reich Railway Management Stettin

Large Oval Postmark type 7.II BArGe reference 7.2.4.6

STETTIN – SWINEMÜNDE: Train number ZUG 546; Trip date 4.8.35; Postmark period 1929-1938

Postcard dated 4.8.35 addressed to Misdroy, West Pomerania. Outbound mail train from Stettin to Swinemünde.

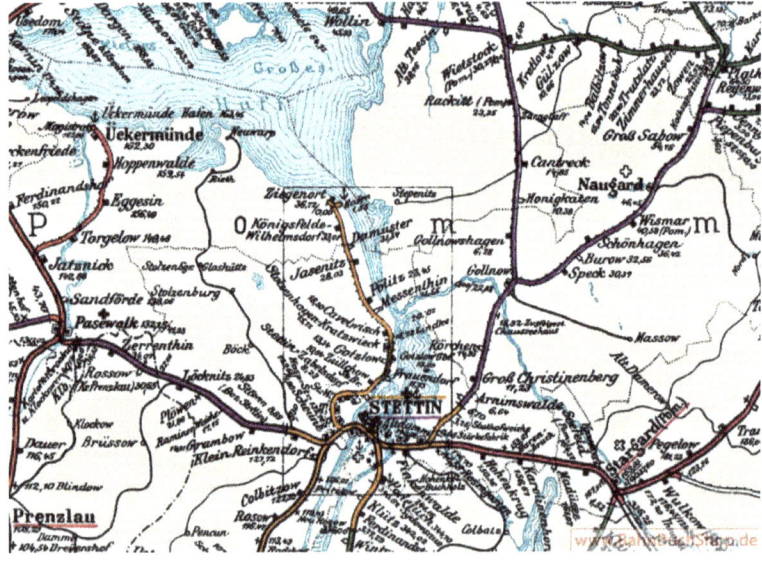

Reichsbahn station Stettin with the original Eastern Railway line crossing the Oder River.

The Route Postmarks Stettin from the BarGe Bahnpost Catalogue

ROUTES	REFERENCE INDEX	PERIOD	NOTES
STETTIN – ALTDAMM	7.1.1.3	1885	
STETTIN – BERLIN	2.1.3.0	1851-1904	
STETTIN – BERLIN	2.1.3.5	1874-1888	
STETTIN – BERLIN	2.1.3.5	1875	
STETTIN – BRESLAU	2	Not issued	
STETTIN – BÜTZOW	2.2.3.0	1874-1886	

ROUTES	REFERENCE	INDEX	PERIOD	NOTES
STETTIN – BÜTZOW	2.2.3.5		1872-1882	Ber.BP 2/75
Stettin – Großziegenort	7.2.1.0		1910-1930	Ber.BP 4/94
STETTIN – HAMBURG	2.2.3.0		1873-1886	Diverse
STETTIN – HAMBURG	2.2.3.5		1880-1883	
STETTIN – HAMBURG	2.2.3.6		1872-1873	
STETTIN – JASENITZ	7.1.1.0		1901-1909	
STETTIN – KLEINEN	7.1.1.0		1886-1890	
STETTIN – KÖSLIN	7.2.1.0		1921-1929	
STETTIN – KÖSLIN	7.2.1.6		1921-1926	
STETTIN – KOLBERG	7.1.1.0		1905-1921	
STETTIN – KOLBERG	7.2.1.0		1916-1943	
STETTIN – KOLBERG	7.2.4.6		1928-1943	Diverse/ F.D.D. 20.7.1928
STETTIN – KOLBERG	7.3.6.0		1933-1936	
STETTIN – KOLBERG (OSTSEEBAD)	7.3.4.1	a	1935-1944	
STETTIN – KOLBERG (OSTSEEBAD)	7.3.4.1	b	1937-1944	
STETTIN – KOLBERG (OSTSEEBAD)	7.3.4.1	c	1936-1942	
STETTIN – KOLBERG (OSTSEEBAD)	7.3.4.1	d	1937-1944	
STETTIN – KREUZ	7.2.1.0		1922-1927	
STETTIN – KREUZ – STETTIN	7.2.1.0		1920-1928	
STETTIN – MISDROY	7.1.1.0		1905-1906	
STETTIN – NEUBRANDENBURG	2.1.3.0		1878	
STETTIN – NEUBRANDENBURG	2.1.3.5		1870-1871	
STETTIN – PASEWALK	2.1.3.0		Without year	
STETTIN – PASEWALK	2.1.3.5		1870-1888	
STETTIN – PASEWALK	7.1.1.0		1888-1911	
STETTIN – STARGARD i. POMM.	7.2.1.0		1921	
STETTIN – SWINEMÜNDE	7.2.1.0		1910-1944	
STETTIN – SWINEMÜNDE	7.2.4.6		1929-1938	
STETTIN – SWINEMÜNDE	7.2.4.6		1940-1943	
STETTIN – SWINEMÜNDE	7.2.6.0		1907-1929	
STETTIN – ZIEGENORT	7.2.6.0		1931-1936	
STETTIN – ZIEGENORT (KR UECKERMÜNDE)	7.3.4.1	a	1933-1938	
STETTIN – ZIEGENORT (KR UECKERMÜNDE)	7.3.4.1	b	1942-1943	

\# F.D.D. - First distribution date

Family Escape in the Rail Mail Car – 1945

31

The following story is a contemporary witness account taken from the diary notes in our family history, which we have preserved for our children and grandchildren.

Dietmar Wicks was born in 1933 in Frankfurt (Main) and grew up with his grandparents. In the autumn of 1940, his grandfather was an employee of the post office in Prussian Holland (East Prussia), and he went to school in the same town. The city was reached by rail via the Elbing-Güldenboden-Maldeuten route. Today it is called Pasłęk and is located in the state of Poland, in the former district of Allenstein. His report reads like this:

"On 20th January 1945, the Soviet armies in East Prussia were advancing from a number of fronts. My grandfather advised to leave the city in the direction of Heiligenbeil in a bus provided by the Prussian Holland post office, which turned out to be a simple rural mail car. Due to the strong onset of winter, the vehicle had to move slowly on the snow-covered roads to Mühlhausen, located between Schlobitten and Braunsberg. Advancing north seemed impossible. The next day we set out on foot with other postal workers and families, keeping a watch in the direction of Elbing, about 28 km away. Many hundreds of refugees joined the trek.

In Elbing we asked at the post office at the train station about getting westbound. Soviet army units commenced firing at the city in a wall of fire. We were advised to travel in a rail mail car; as we remember, it was the last train leaving Elbing. In this rail mail car, the only rail mail attendant had built a makeshift 'bed' in the skylight arm. As far as I can remember, there were as many as 60-80 refugees in the mail vehicle. We were accommodated in the protective compartment for sitting, which was actually 'overcrowded' with 15-20 people. Stretch your legs, impossible! The train left Elbing at around 11 p.m. and after several hours of stopping it reached Dirchau two days later, about 46.5 kilometers away by rail.

The train was barely heated, there was no radiator in the protective compartment of the rail mail car, windows and outer doors froze up, bitterly cold carriages to overcome for the entire 'refugee crew'. We didn't have anything warm to eat. When we stopped for a long time in the open, the snow was used for washing. On the way to Dirschau, through the Tucheler Heide forest, the coal of the locomotive was running low. We collected wood for the machine from the snow-covered forests.

After about eight days in Bütow (Pomerania) we left the lifesaving and 'gas-friendly' rail mail car and drove with trains that were still in service via Zollbrück to Stolp. My grandfather's post office was outsourced here. After the Soviet army broke through in Pomerania, East Pomerania and Danzig were cut off from the Reich. The people tried to leave the area by sea. My grandparents and I got as far as Danzig (now Gdansk). We experienced the pillage and the executions by the Russian soldiers here with all the atrocities (but that is another chapter).

My grandparents wanted to return to Prussian Holland after the fighting in Danzig. Some of the refugees believed that the Soviets could be pushed back from East Prussia. On 1st April 1945, two old people and a twelve-year-old boy set out from Danzig via Praust and Dirchau. We set off in a southward direction along the Vistula to Munsterwalde, as the bridges were destroyed. Then we continued via Marienwerder, Finckenstein, and Christburg to Holland. On the orders of the Polish authorities, we had to leave the city for the second time in an unheated freight car in November 1945. After a stopover on the island of Rügen, I moved to Northeim am Harz in Lower Saxony."

Dietmar Wicks found employment as a post office boy on 1st April 1947 at the post office in Northeim. After an apprenticeship, he wrote his examination in 1950 at the Head Post Office in Braunschweig. Thus ended his hope of working his former Prussian Holland district, a German town no more. As a rail mail attendant for the Northeim post office,

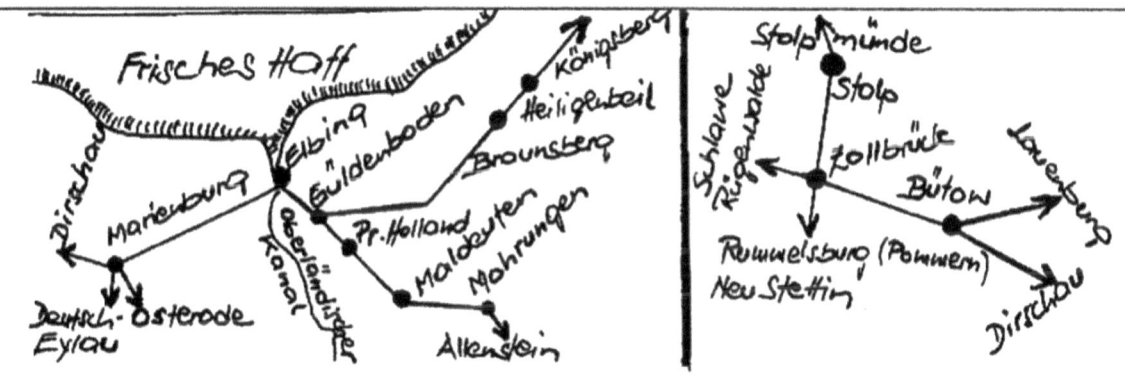

Route of the mail train from Allenstein – (Pr.) Prussian Holland – Elbing – Marienburg – Dirchau – Zollbrück – Stolp

he drove on the Northeim-Walkenreid, Ottbergen and Duderstadt rail mail routes. In November 1954, he was transferred, at his request, to the railway post office (BPA) 19 in Frankfurt (Main).

After his rail mail service and service in the route department of the BPA 19, he retired in 1996. A postler's life that began in 1945 in a frozen rail mail car ended for Dietmar Wicks in 1996 with a last trip in a modern, warm rail mail car on the Basel SSB-Frankfurt (M) route.

Withdrawal, Surrender and Dissolution of Reich Railway Management Stettin

32

Headquarters of the Stettin Post office. Used postcard

The front divisions of the Red Army commenced an attack on the Vistula on 12 January 1945. The military movements were continued by the 1st Belorussian Front under Georgy Zhukov and the 2nd Belorussian Front under Konstantin Rokossovsky. The military formations advanced to the Oder and the Baltic Sea in the Pomeranian region. In East and West Prussia, the manoeuvres caused the railways to constantly alter their course of retreat, whilst the German II Army found its exit route to the west cut off.

Zhukov's armies advanced through the Reich Railway management areas of Posen, Danzig and West Prussia to reach the River Oder to the north of Küstrin and established a bridgehead on the western side less than 60 km from Berlin on 31 January 1945. The Russian 5th Army at Zehden, West Pomerania, under the command of Roszkowski, pressed against the newly formed German defence group "Weichsel" under Himmler, with the 11th Army and 3rd Panzer Army forming a broad front, heading north to the Baltic Sea on 6 February 1945.

Before the land battles began, in contrast to the other eastern Reich Railway management areas, the district of Reich Railway Management Stettin was constantly under aerial attack. The United States Air Force and RAF bomber squadrons stationed in the United Kingdom launched numerous attacks on the district's major towns and cities. Thus, between 26 August 1940 and 6 October 1944, Stettin was exposed to 19 major air attacks that largely destroyed the city and also affected the railway facilities.

The ongoing air raids on Berlin also led to aerial war damage in the suburbs, and the rocket test station in Peenemünde was attacked on 17 and 18 August 1943 and on 6 October 1944.

From February 1944, airstrikes continued on a large hydrogenation plant (synthetic fuel plant) in Pölitz, West Pomerania, close to Stettin. On 6 October 1944, Stralsund was heavily bombed, and on 12 March 1945, there was a large-scale attack on the harbour area of Saßnitz. Then Świnoujście, a port city in Pomerania, was bombed on 7 March 1945. The Red Army continued to advance all around Pomerania. The district increasingly came under fire from Russian aircraft, while in February 1945, in the western districts, Montgomery's forces advanced with the assistance of the Allied bomber squads.

By the end of January 1945, the Red Army had advanced into the area surrounding the eastern cities of RBD Pomerania and came to a halt on the outskirts of Reich Railway Management Stettin. On 20 February, the Red Army managed to reach a line north of Konitz - Nystatin - Falkenburg - Arnswalde - Pyritz - Oder and thus penetrate the district. On 27 February, the station in Nystatin was occupied, while the city fell into Russian hands the following day.

The concentric northward Russian advance interrupted the main route between Schlawe and Köslin, and eastern Pomerania was cut off from the rest of the Reich. The retreating German army from East and West Prussia reached the Baltic Sea at Labus. As East Pomerania was occupied by the Red Army, the railway management areas were no longer managed by Stettin. Only a few days remained before the Germans lost possession of Danzig and the Russian occupation of the east was complete. The Russian forces, now facing east, occupied the town of Bütow without a fight on 7 March. On 8 March they took Stolp and on 10 March, Lauenburg. Thus, the entire Reich Railway Management (RBD) Danzig area of Eastern Pomerania was in Russian hands. Further west, the troops of Konstantin Rokossovsky, and some of Zhukov's, advanced concentrically to the north and occupied Naugard and Stargard without any significant fighting. On 5 March, Belgard and Köslin were evacuated before Russian forces captured the two cities.

On 7 March, Kolberg and Stettin were under siege. With the evacuation of the Stettin district of Gollnow, the Red Army occupied the town on the same day. General Hasso von Manteuffel, a German Baron, who had been in command since 10 March, was unable to defend the bridgehead after March 19. He evacuated most of it and, on 20 March, had the Oder bridges blown up. On 25 March 1945, Stettin was abandoned.

Klaus Raatz, a railway mail collector and retired Deutscher Bahn driver who witnessed the build-up and the fighting, reported that on 5 March, the belated evacuation of Schivelbein was underway with little hope of progress. On one of the planned evacuation trains, the locomotive was shot to pieces by Russian aircraft. In front of it was a second train intending to depart with evacuees. The planes reappeared and shot this locomotive too, but an attempt to bomb the third locomotive failed. However, the train couldn't leave due to line damage, and later in the day, Russian tanks rolled in from Labes and reached the district of Schivelbein. The route between Schivelbein and Belgard had meanwhile been reached and interrupted by an armoured battalion that had penetrated east of Schivelbein.

The purpose of these individual tank movements was the formation of a double-encircled area by the 19th, 3rd and 2nd Guards Tank Army in Labes, Bad Polzin and Schivelbein. A refugee train from the Schivelbein area departed eastwards along the Bad Polzin-Gramenz route but changed direction onto the Gramcnz-Belgard route, which was still open at that time. The train drove directly into the path of the advancing Russian armoured spearhead.

On 4 March 1945, the Belgrade-Kolberg rail line was blocked in a driving snowstorm and a train was stranded near Kolberg. As the Russian air raid on Kolberg commenced at 4.30 p.m., the Kolberg station entrance sustained a direct hit and trains were unable to enter the station. In contrast, on the Kolberg-Köslin route, trains were engaged in return trips for the benefit of commuter traffic for a few more nights. Since Köslin provided an area relatively free of fighting, on the nights of 7 and 8 March, the Kolberg train ran shuttle trips, moving masses of refugees from all around the Baltic Sea to a safer area. In the fierce fighting in the station area, the district changed hands several times. Eventually, all that was left of the locomotives was debris. On 18 March 1945, Kolberg was overrun by the Red Army.

The Red Army units reached the eastern parts of Stettin on 8 March 1945, and the city was cleared according to plan. The peak of the railway operations was around 10 March at Altdamm and the town of Podejuch on the right bank of the Oder. These towns were occupied by Soviet troops of the 1st Belorussian Front on 20 March 1945. On March 18, when all the Oder bridges were blown up, the main line to Podejuch was the only

Prussian-built T 16.1 freight locomotive from Stettin

connection remaining through the Grafscharte Mountain to Stettin; the main station was the only means of escape to the port facilities.

Regarding evacuation from Stettin main station, rail lines leading from the city were for the most part reduced, and all transportable material was transported to the safety of the west, mostly to Schleswig-Holstein. With the exception of some fixed assets, most of the surviving Prussian-built T 16.1 locomotives and all movable rolling stock were driven west.

However, this did not apply to much of the district east of the Oder, although it had been initially successful. Much material and many trains were transferred to the west by the remaining Reich Railway Management (RBD) staff. The old and abandoned Pomeranian main line running through Danzig, Stolp, Stargard and Stettin, including those routes connecting Berent to Bützow, Rummelsburg, Neustettin, Falkenburg and Ruhnow, was used to evacuate RBD assets from Danzig and refugees from the surrounding areas Danzig city to the west.

Due to a lack of coordination, these single-rail routes created a total blockage near stations. The entire operation was mired in inefficiency, and Stargard was virtually paralyzed, with blocked engines inhibiting the smooth running of any plausible plan. Many trains had to stop or even be driven back. On 28 February, the 19th Army captured Neustettin, West Pomerania, and the battle of Stargard took place on 4 March.

Most of the equipment still to be moved west of the Oder remained east of the river, much of it in Stargard railway marshalling yards and the RBD offices east of the Oder. An evacuation by sea as in other areas could not take place due to the lack of exit lines to ports; the few available shipping ports and the surrounding area had to be used for the evacuation of the population and the wounded. Nor was it possible to bring railway material west to any significant extent via the Ostswine-Świnoujście ferry route which escaped serious damage until 12 March 1945, when an air raid destroyed 55% of the city buildings and sank several dozen ships. The seaport city was finally overcome on 5 May 1945, and it was the last city in Pomerania to fall. On 20 March 1945, the RBD Stettin moved its headquarters from Stettin to Neustrelitz, Mecklenburg-Vorpommern. In the bunkers of Stettin Main Station, an operational command post of the RBD was maintained.

At the end of March 1945, the Stettiner Main Railway Station came under artillery fire. Three passenger trains had departed and were re-routed along minor railway routes: Scheune - Zabelsdorf - Pölitz. From Zabelsdorf, Brandenburg, they re-joined scheduled services operating to Berlin, Hamburg and Stralsund. All other removals from Stettin RBD offices took place during this time using freight trains.

On 16 April 1945 at 5.00 AM, the 1st Belorussian Front commenced their attack and opened a new Russian front around the RBD Stettin with ten battalions, including three armoured brigades. It was not until 18 April that a larger bridgehead was created between Seelow Heights and the Finow Canal, from which a concentric attack was made on Berlin over a three-day battle. By 25 April, the entire southern district to the Finow Canal was in Russian hands.

The 2nd Belorussian Front under General Roszkowski succeeded in erecting bridgeheads over the Oder around Hohenzehden on 15 April and near Pölitz (West Pomerania) on 17 April. The 2nd Belorussian Front arrived between Schwedt and Stettin on 25 April with four army divisions and broke through the 3rd Panzer Army's line around the bridgehead south of Stettin, moving rapidly westwards. On 26 and 27 April, there was the non-combative evacuation of the rest of Stettin. On 29 April, Anklam, Mecklenburg, was occupied, and on 30 April, Demnin, Tribsees and Greifswald were handed over without a fight. On 2 May, the town of Barth, Mecklenburg, fell, and on the same day, the islands of Usedom and Wollin came into Russian hands. On 1 May, Stralsund was occupied, following the surrender of the city by a businessman, Gerhard Poggendorf, who raised the White Flag.

Because of these events, the last railway workers left Stettin on the evening of 26 April, and the RBD relocated to Neustrelitz. Mecklenburg moved to Greifswald on the same day. On 29 April, the command train moved to Ribnitz Bahnhof, Mecklenburg, a safer distance away. Staff numbers were rearranged, leaving most of the staff, including the redundant sections of the RBD, having to resettle in Mecklenburg. Some of the operational files were not destroyed, just left behind.

On the evening of 29 April, the command train arrived in Hornstorf Station near Wismar and on 1 May was able to continue through to Lübeck, departing around 4:00 p.m.

The train with its staff continued through Teufelsberg (Devil's Mountain) and the Grunewald forest in Berlin and connected with the Malente-Gremsmühlen-Lütjenburg route in Schleswig-Holstein. This was the only safe destination; it was not possible to be stationary for too long with the Russian forces closing in and lines to Neustadt declared closed.

Under the orders of the Reich Ministry of Transport, on 2 May 1945, the RBD Stettin was officially dissolved. The command train was moved to a newly formed settlement centre on the outskirts of the small forest town of Benz in Maleate and was joined by the command train for the RBD Schwerin. They were finally stationed in Malente Gremsmühlen. After the war, the two trains in Malente came under the control of the settlement office of the RBD Hamburg, which was dissolved at the end of November 1945.

The Prussian S 6 (later DRG Class 13.10–12) was a class of German steam locomotive with a 4-4-0 wheel arrangement operated by the Prussian state railway for express train services on the Eastern Railway between Stettin and Eydtkuhnen. The Head of the Locomotive Design and Procurement Department, Robert Garbe, proposed to the Locomotive Committee in 1904 a design by Linke-Hofmann of Breslau for a 4-4-0 superheated, express train locomotive. This was an evolutionary development of the Prussian Class S 4. In 1905/1906 Garbe pushed through the construction of the S 6. Between 1906 and 1913 a total of 584 units were manufactured.

Railway Station Postmarks – Deutsche Reich Period 1889–1945

33

Frame Cancel Railway Station Office Postmark

STARGARD i. POM.BAHNHOF; Dispatch date 25 6 72

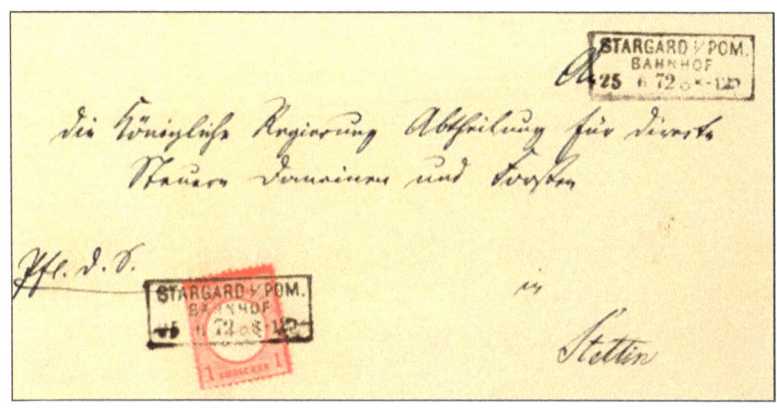

Letter received at Stargard station for onward transport was postmarked with the station office stamp addressed to head post office in Stettin. Deutsche Reich Inland Letter rate paid.

Circular Postmark

STETTIN BAHHPOST; Dispatched date 2.10.97; Postmark period 1872-1915

Characteristics of the Postmark: Locket above the word "Bahnpost"; no data bridge; Time below data line; Station name in the upper part.

Railway Mail at Frankfurt Oder 34

The first proposals for the construction of a railway between Berlin and Frankfurt (Oder) were made in the 1830s. August Leopold Crelle was largely responsible for the planning. The location of the Frankfurt (Oder) station was not decided until after construction of the line began. It was built in a y shape. The terminal station on the brickyard hill (Ziegeleiberg) just outside town was renamed Briesener Straße in recent times. A location nearer the town was not possible because of the large differences in height in the Oder valley. On 23 September 1842, the station and the line between Berlin and Frankfurt (Oder) were opened.

The Niederschlesisch-Märkische-Eisenbahngesellschaft (Lower Silesian-Markish Railway Company, with "Markish" referring to the March of Brandenburg/Mark Brandenburg) was founded in 1842. Its goal was the extension of the line from Frankfurt (Oder) to Breslau. The line was opened to Breslau on 1 September 1846. The Prussian government built the Prussian Eastern Railway (Preußische Ostbahn) to Königsberg, and the station at Frankfurt (Oder) was one of its principal stations along the eastern network.

The city, the region and the railway were exposed to violent destruction at the end of the war when Soviet troops occupied the city on 23 April 1945.

Line Stamp Postmark type II BArGe reference 2.2.3.5

BERLIN – FRANKFURT. ODER; Tour/Trip V R; Trip date 17/ 3; Script common Roman; Departing station / Data line / Arrival station; Postmark period 1883-1887

Pre-paid Reich postcard dated 18.3.87 from Berlin to Zabrze, a western district of the Silesian Metropolis. Mail train outbound from Berlin Stettiner Station to Frankfurt (Oder), trip V R (Return). (Cover from the archives of the Bundesarbeitsgemeinschaft e.V. Frankfurt)

c.1900 Frankfurt a.O. Kaiserliche Post office. Unused postcard.

The G 8.1 on The Eastern Railway main line. The engines were designed by Robert Garbe and built between 1913 and 1921, forming the largest class of state railway locomotives in Germany. The boiler was larger than that of the G 8, and the driving wheel diameter was 1,350 mm. The locomotive was designed to be heavy enough to haul even the heaviest loads without sanding, due to its higher adhesive weight. Because it had a 17.6 t high-axle load, the G 8.1 could, however, be used only on main lines. In addition to its employment for heavy goods traffic, it was later used for heavy pusher duties as well.

SMALL AND LARGE OVAL POSTMARKS OF FRANKFURT ODER

Small Oval Postmark type 7.I BArGe reference 7.1.1.0

FRANKFURT (O)-MESRITZ- POSEN; Train number ZUG 272; Trip date 12 6 00; Postmark period 1893-1917

Postcard, with a receipt postmark 14 JUN 1900 in Ludwigsburg, mailed from the district of Posen; mail was collected at Meseritz, the train returning to Frankfurt (Oder) station.

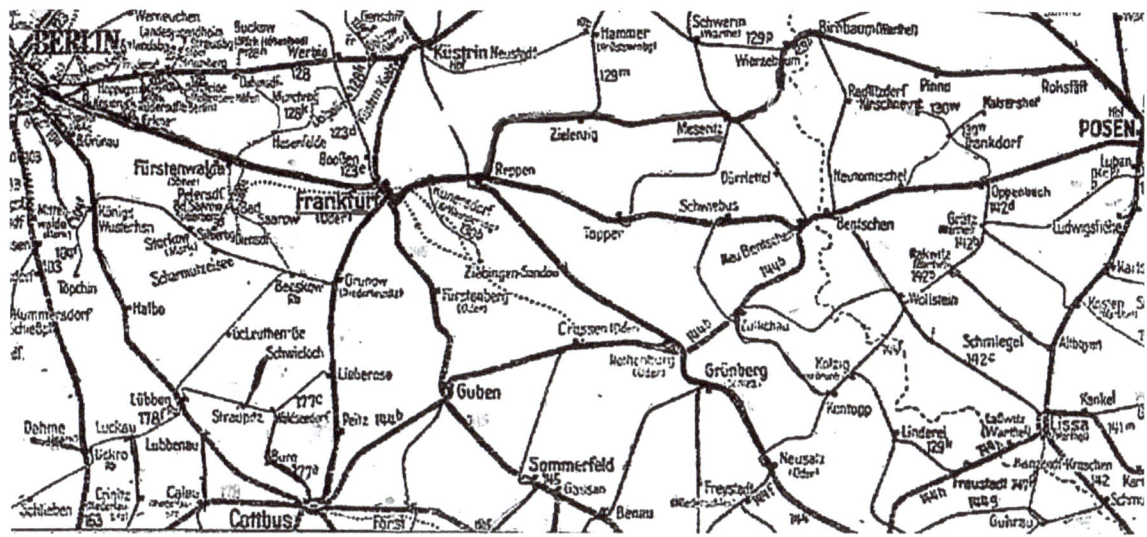

Map supporting the above route postmark: Frankfurt (Oder) – Meseritz-Posen

Leopold Crelle was largely responsible for the planning. It has been one of the most important railway stations in the German state of Brandenburg since 1945. It was the new border station for transport to and from Poland, which was specified in the Potsdam Agreement.

Frankfurt (Oder) Rail Station

Small Oval Postmark type 7.1 BArGe reference 7.1.1.0

FRANKFURT (ODER)- BERLIN; Train number ZUG 444; Trip date 2 9 10; Postmark period 1888-1919

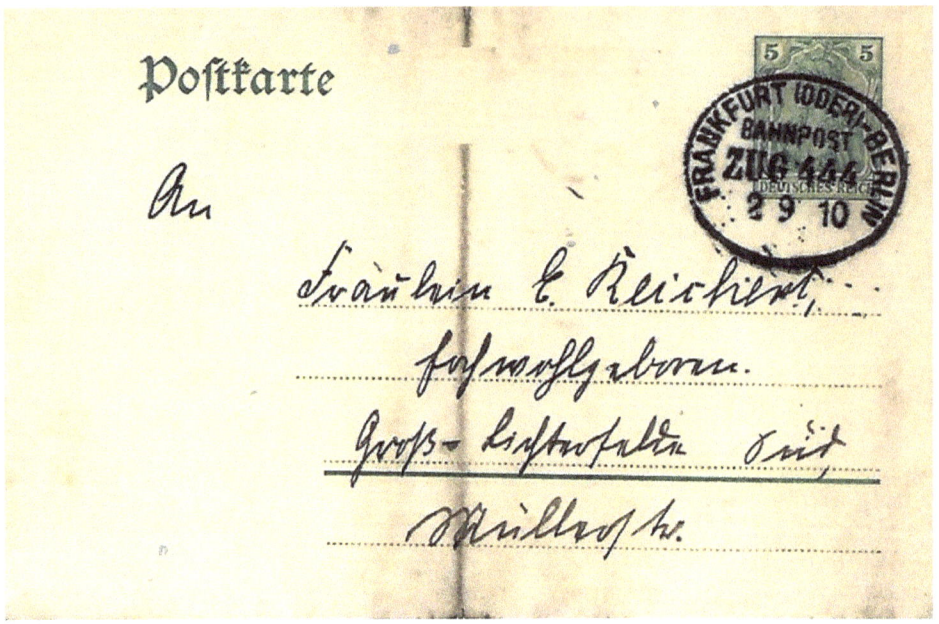

Mail train returning from Berlin to Frankfurt (Oder) station

RAILWAY CONNECTIONS BETWEEN EASTERN RAILWAY STATION FRANKFURT (ODER) AND

Once the railway lines from Berlin via Frankfurt (Oder) to Guben were connected, they continued to Breslau. Railway connections from Frankfurt (Oder) continued to the northeast in the direction of Küstrin-Kietz. Then there were the lines running eastwards and in a southerly direction, crossing the deep Oder valley to the Warsaw station. The station at Eberswalde had been a junction since 1866 when a line to Wriezen was opened. The planners then proposed, for economic reasons, to connect the line to Frankfurt (Oder) in 1876/77. The Prussian connecting line from Priestewitz to Frankfurt (Oder) via Cottbus was completed after the network became nationalised on 1 September 1883. By early 1900, Frankfurt (Oder) was an important junction, with multiple routes converging.

The Großenhain-Priestewitz railway was a single-track electrified main line in the German state of Saxony. It was originally built by the Großenhain Branch Railway Company (Zweig-Eisenbahngesellschaft zu Großenhain) but was nationalised by the Prussian state. The line was later extended from Cottbus to Frankfurt (Oder), providing those who travelled from Sachsen a connection to the Eastern Railway network.

Small Oval Postmark type 7.I BArGe reference 7.1.1.0

FRANKFURT (O) - PRIESTEWITZ; Train number ZUG 406; Trip date 26 5 91; Postmark period 1888-1937

Pre-paid cover from Peitz, south-east of Brandenburg, dated 26 May 1891 to Dresden, with a city of Dresden receipt postmark 27 5 91. Mail train outbound from Frankfurt (O) to Priestewitz.

Small Oval Postmark type I BArGe reference 7.1.1.0

FRANKFURT (O) - PRIESTEWITZ; Train number ZUG 975; Trip date 21 9 30; Postmark period 1888-1937

Post-paid postcard from Priestewitz addressed to Berlin. Mail delivered to Priestewitz station on return mail train. The ZUG 975 departed Priestewitz 14.19, arrived at Frankfurt (Oder) 20.08; transferred to Berlin-bound train as bag mail for final journey (Reich time table).

Mail trains from Berlin to eastern destinations would almost always stop at the Frankfurt (Oder) station. The following trips fell into two categories until 1945.

104 HISTORY OF THE EASTERN RAILWAYS CONSTRUCTION AND EXPANSION

1) Local traffic and regional trains stopped at Frankfurt (Oder) going towards Breslau.

Berlin-Charlottenburg → Berlin Schlesian Main Station → Fürstenwalde (Spree) → Berkenbrück (RBD Osten) → Briesen (Mark) → Jacobsdorf (Mark) → Pilgram → Rosengarten → Frankfurt (Oder) to Guben and Sommerfeld.

2) Express D ZUG trains crossed Frankfurt (Oder) in the direction of Eydtkuhnen

Berlin-Charlottenburg → Berlin Schlesian Main Station → Frankfurt (Oder) to Dirchau and Eydtkuhnen.

Railway Time Table for the Frankfurt (Oder) – Priestewitz route (See page 103):

Railway Postmarks of Frankfurt (Oder) to look for

Large Oval Postmark type III BArGe reference 7.3.2.0

BERLIN – KÖNIGSBERG N.M.; Train number Z. 686 ---; Trip date 4.4.31; Postmark period 1921-1944

Pre-paid postcard with the Upper Silesian coat of arms written from Königsburg N.M., dated 3.4.31 and addressed to Berlin. Mail transport returning from Königsberg N.M. to Berlin. Postcard Inland postage rate on Michel Ganzsachen P 189

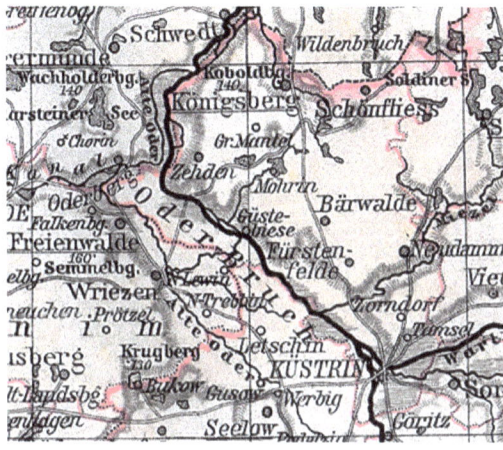

The district of Königsberg Nm. and Neumark in the province of Brandenburg: the eight cities of Bad Schönfließ, Bärwalde Nm, Fürstenfelde, Königsberg Nm., Küstrin, Mohrin, Neudamm and Zehden (Oder), as well as 99 other municipalities and two forest state districts. In the spring of 1945, the district was occupied by the advancing Red Army.

Large Oval Postmark type 7.III BArGe reference 7.3.6.0

BBERLIN - EBERSWALDE - FRANKFURT (ODER); Train number Zug 448; Trip date 4.4.38; Postmark period 1928-1941 Diverse

Postcard dated 4.VI.38 delivered to Berlin on a mail train returning from Frankfurt (Oder)

Large Oval Postmark type 7.III BArGe reference 7.3.6.0

BERLIN - FRANKFURT (ODER); Train number Zug 12-2; Trip date 15.11.75; Postmark period 1971-1979 Diverse

Letter addressed to Hannover departed on a return mail train from Frankfurt (Oder) to Berlin.

The Route Postmark Catalogue for Frankfurt Oder

THE ROUTES	REFERENCE	INDEX	PERIOD	NOTES
FRANKFURT A/O – ANGERMÜNDE	2.2.3.0		1880-1881	
FRANKFURT A/O – BENTSCHEN	3.1.2.7	a	1880	
FRANKFURT A.O. – BENTSCHEN	3.1.2.7	b	1883	
FRANKFURT (O.) – BENTSCHEN	7.1.1.0		1888-1905	
FRANKFURT (ODER) – BERLIN	7.1.1.0		1888-1919	Diverse
FRANKFURT A.O. – CÜSTRIN	2.1.3.0		1871-1874	
FRANKFURT (ODER) – DRESDEN	7.1.1.0		1914-1932	
FRANKFURT (ODER) – DRESDEN	7.3.4.0		1938-1944	
FRANKFURT (ODER) – EBERSWALDE	2.2.3.0		1883-1888	
FRANKFURT (ODER) – EBERSWALDE	2.2.3.9	a	1883-1887	
FRANKFURT (ODER) – EBERSWALDE	7.2.1.0		1929-1935	
FRANKFURT (ODER) – EBERSWALDE	7.2.1.6		1932-1934	
FRANKFURT (O.) – MESERITZ – POSEN	7.1.1.0		1893-1917	WV.PL to 1918
FRANKFURT A.O. – PRISTEWITZ	3.1.2.6	a	1881	
FRANKFURT A.O. – PRISTEWITZ	3.1.2.6	b	1883	
FRANKFURT A.O. – PRISTEWITZ	3.8.2.0	a	1881-1886	
FRANKFURT A.O. – PRISTEWITZ	3.8.2.0	b	1880-1886	
FRANKFURT (O.) – PRIESTEWITZ	7.1.1.0		1888-1937	
FRANKFURT (ODER) – PRIESTEWITZ	7.1.1.0		1907-1930	
FRANKFURT (ODER) – PRIESTEWITZ	7.3.1.0		1930	
FRANKFURT (ODER) – PRIESTEWITZ	7.3.6.0		1933-1943	
FRANKFURT (ODER) – SCHWERIN	5.0.0.0		1961-1962	
FRANKFURT (ODER) – STETTIN	7.3.6.0		1934	

c.1900 Cüstrin North Railway station rebuilt 22 October 1861. Unused postcard

STEAM ENGINES OF THE DEUTSCHE REICH PERIOD AND RAILWAY STATIONS

¾ coupled express train locomotive S 5 class of the Prussian State Railway. The locomotive was built by the Vulcan machine factory in Stettin in 1898. The gross weight of the locomotive was 533 300 kg. Speed 100 km/h on the Eastern Railway routes. The locomotive was initially very similar to its French models. In later modifications, the French driver's cab was replaced by a typically Prussian one, and the boiler superstructures were brought into line with Prussian standards. The locomotive was retired in 1916.

c1937 Stargard railway station. An unused postcard.

Railway Construction in and around Kreuze Station 35

The station at Kreuz as seen in early 1900

In the 1840s, land was purchased for the Stargard-Posen railway, which was opened in 1848. The first section was called Dragemünde. In January 1846, the Berlin-Küstrin-Driesen-Schneidemühl-Thorn-Graudenz-Dirschau line was approved, and the construction of the railway line began.

In 1847, the work stopped due to worker unrest that broke out in Berlin. It was not until 1848 that the state parliament approved the continuation of the railway line. The first train on the newly built railway line was lowered from Stettiner Main Station in Berlin on

26 July 1851 at 11.00 a.m. The first train departed for Bromberg, and another departed in the early hours of 27 July (at 02:30 a.m.) from Bromberg to Berlin.

Kreuz (on the Ostbahn) was part of Prussia. When the service commenced, it was soon recognised as an important railroad junction, with two major lines crossing there - the Berlin-Bromberg and Posen-Stettin connections. The town owes its existence to the rail, as it developed only after 1848 when the Posen-Stettin line was opened.

The Kreuz station, with the southwest-northeast orientation of its tracks, was designed from the beginning to be linked to the Eastern Railway from Berlin to the Lower Vistula and East Prussia. The Stargard-Posen Railway crossed the Kreuz station on 27 July 1851.

From 12 October 1857, the line from Küstrin to Kreuz was also in operation, so two important railway lines crossed at Kreuz (which led to the name "Kreuz"). The station and the town of Kreuz was the responsibility of the Lukatz municipality under administrative law.

The train first ran from Berlin via Stettin and Kreuz to Bromberg and back. Connecting trains ran from Kreuz to Posen. In 1895-1896, the Kreuz station received a further connection to Rogasen, which branched off at Dratigmühle from the Posen railway line to Filehne and had a branch line to Czarnikau.

On 1 December 1899, the narrow-gauge Kreuz-Schloppe railway line was completed, and in 1904 this route was further expanded beyond Schloppe to Deutsch-Krone. In 1910, the facilities of the Kreuz train station (passenger and marshalling yard, as well as a workshop) were adapted to the requirements of modern rail traffic. Apart from Schneidemühl, Kreuz was the most important operating centre on this part of the Eastern Railway.

The town of Kreuz was shaped mainly by railroaders and merchants. Water handling was possible via the network port from 1912.

In 1935, the Kreuz-Guben railway construction through Driesen was completed due to the demarcation of land and a new border between Germany and Poland. During the Greater Poland/Posen uprising in 1918/19, there were supposed to have been fierce battles for the Kreuz railway junction. According to the stipulations of the Versailles Treaty, the border between Germany and the newly created Republic of Poland came into effect. Due to these changes, in 1935 it became necessary for the construction of a railway line between Kreuz and Guben via Driesen.

Are you interested in trains and railways? If the answer to either of these questions is yes, then the Railway Philatelic Group (RPG) is the place for you. A quarterly journal is published, and an annual general meeting is held where all collectors have an opportunity to showcase their collections. For further details, have a look at their website **www.railwayphilaticgroup.co.uk**. Collectors interested in pursuing railway philately can join our Facebook page "**Railway Philatelic Group**" and post covers and cards.

Railway Postmarks of Cüstrin Station 36

7.2.0.0 Large Oval Type II

Data bridge with "Z", Train number, Day, Month, Year between horizontal lines, Decorative mouldings below: Variation of "Z", "Zg" or "ZUG", also "Z.", "Zg." or "ZUG."

Other Characteristics

- 7.2.-.1 Index behind Route or BAHNPOST
- 7.2.-.2 Index behind Data Bridge
- 7.2.-.3 Index in decorative moulding
- 7.2.-.4 Star before Index behind route or
- 7.2.-.5 Star behind Data line
- 7.2.-.6 Others in decorative mouldings (s. side 7-6)
- 7.2.-.7 Roller press postmarks with wave lines
- 7.2.- 8 Others in decorative moulding (s. Side 7-6); Index or Star behind Route or BAHNPOST

RAILWAY POSTMARKS OF CÜSTRIN STATION   113

Large Oval Postmark type 7.II BArGe reference 7.2.1.0

CÜSTRIN - FRANKFURT (ODER): Train Number Z. 970; Trip date 12.11.25; Postmark period 1917-1929

Post-paid letter from Pomerania to Westfalen. Mail train outbound from Cüstrin – Cüstrin Altstadt – Cüstrin Kietz – Reitwein – Podezig – Lebus – Kllestow (Kr.Lebus) to Frankfurt (Oder).

Large Oval Postmark type 7.II BArGe reference 7.2.4.6

KÜSTRIN - STARGARD (POM.): Train number ZUG 0904. ; Trip date 1.3.43; Postmark period 1930-1943

Postcard 1.3.43 addressed to Berlin. Mail train ZUG 904 on a return trip, departing Stargard 13.20; arrival Küstrin 16.13

Railway Time Table for Küstrin

129b Küstrin Neustadt – Glasow – Arnswalde / Pyritz – Stargard (Pom)

Alle Züge 2. 3. Klasse

[Timetable image — detailed train schedule table with columns for train numbers (921, 903, W1521, 927, 917, 907, 1523, 1525, etc.) and station stops including Küstrin N.Hbf, Wilkersdorf-Zorndorf, Zicher, Neudamm, Berneuchen, Ringenwalde (Neum), Rosenthal (Neum), Rostin, Soldin, Glasow, Ernestinenhof, Adamsdorf, Chursdorf (Neum), Dieckow, Berlinchen, Bernstein, Blankensee (Pom), Alt Libbehne, Sommenthin, Arnswalde, Lippehne, Mellentin, Naulin, Pyritz, Groß Rischow, Friedrichsthal (Pom), Groß Schönfeld, Warnitz-Damnitz, Klützow, Stargard (P).]

Prussian Eastern Railway Station Schneidemühl

37

The Schneidemühl main station was built between 1851 and 1876 as part of the Royal Prussian Eastern Railway. The station was rebuilt when it was extended in the 1920s.

Frame Cancel Railway Station Postmark

SCHNEIDEMÜHL BAHNHOF: Dispatch date 20 1 69

Letter dispatched from Schneidemühl rail station addressed to Bromberg

Schneidemühl main station.
Postcard postmark 1.6.04

CIRCULAR RAILWAY POSTMARKS SCHNEIDEMÜHL

Large Oval Postmark type II BArGe reference 5.2.3.2

SCHNEIDEMÜHL - CONITZ; Train/Trip Number II; Trip date 6 12 72; Postmark period 1872

1/3 Groschen MiNr.17

Large Oval Postmark type 5.II BArGe reference 5.2.3.2

SCHNEIDEMÜHL - FLATOW; Train/Trip number I; Trip date 2 3 73; Postmark period 1871-1873

Pre-printed postcard dated 1873 from Konitz addressed to Thorn; supports a period of political transition. Mail train departed Flatow on a return trip to Schneidemühl.

SMALL OVAL POSTMARKS OF SCHNEIDEMÜHL 1889-1939

Small Oval Postmark type 7.I BArGe reference 7.1.1.0

SCHNEIDEMÜHL - STARGARD: Train number ZUG 758; Trip date 10.4.15; Postmark period 1896-1925

Small Oval Postmark type 7.I BArGe reference 7.1.2.0

BERLIN - SCHNEIDEMÜHL: Train number ZUG 346; Trip date 16 12 23; Postmark period 1922-1936

Official postage issue
MiNr.85

LARGE OVAL POSTMARK OF SCHNEIDEMÜHL 1910-1945 AND STATION BELGARD

German Belgard was once a historic town in Middle Pomerania. It was, until 1920, the most important railway junction of Middle Pomerania, linking Kolberg and Danzig with Stargard.

The first post office in Belgard was opened in 1825. In 1858 the first railway connecting Belgard to Köslin and Schivelbein was completed. It was extended to Stargard and Neustettin in 1878. Belgard became part of the German Empire in 1871.

In the final weeks of the war, the Red Army occupied the town on 4 March 1945. According to the terms of the Potsdam Conference, the town was returned to Poland.

Large Oval Postmark type 7.II BArGe reference 7.2.1.0

SCHNEIDEMÜHL - BELGARD: Train number Z. 742; Trip date 25.7.34; Postmark period 1923-1945

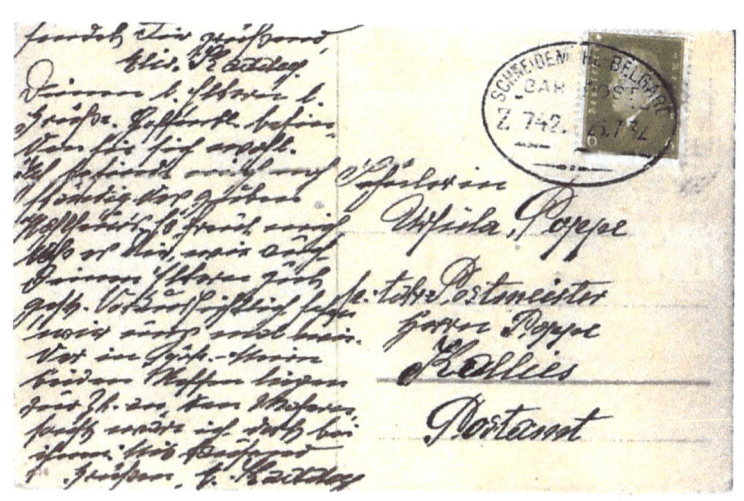

Post-paid postcard addressed to the Postmaster at Kallies Post Office i.pom. Mail train Z. 742 returning Belgard departing 5:47 a.m. Arrived Neustettin 7:27 final stop. Used a postmark Schneidemühl to Belgard. Mail was delivered from Neustettin to Stargard and Kallies by automobile.

The route stops for mail train ZUG/Z.742:

Schneidemühl - Lebenhnke - Plietnitz - Linderhof - Betkenhammer - Jastrow - Ratzebuhr - Bahrenbusch - Lottin - Thurow (Kr. Neustettin) - *Neustettin* - Dallenthin - Eschenriege - Elfenbusch - Gramenz - Villnow - Groß Tychow - Klefheide - **Belgard** - Körlin - Dassow - Fritzow - Jaasde - Degow - Alt Tramm - Kolberg (final destination of the train).

Rail time table for Zug 742 Belgard – Neustettin (See page 117)

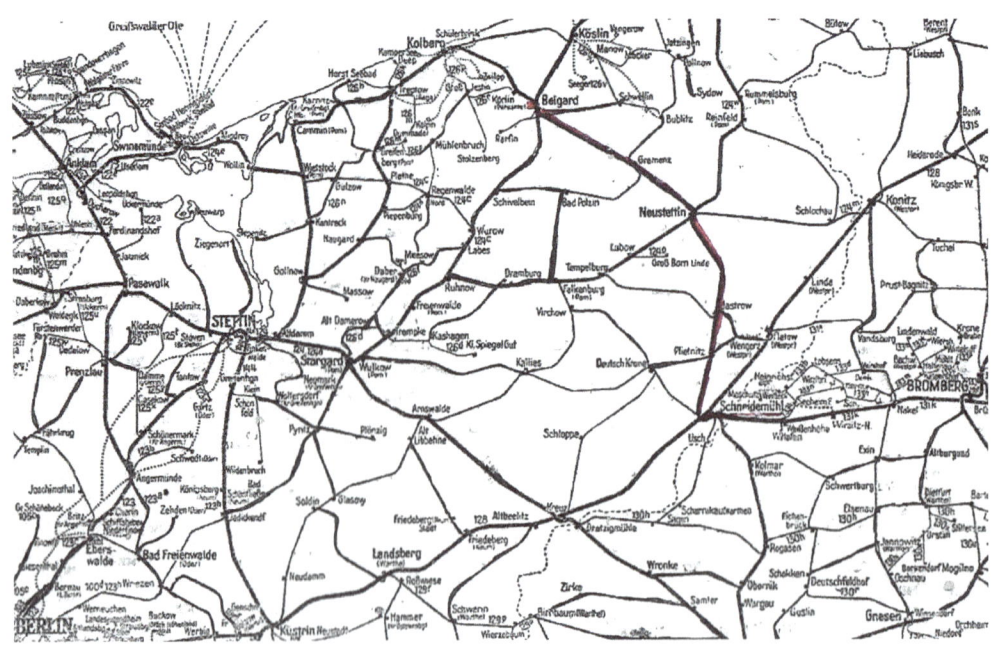

The Route Map of Pomerania Belgard – Neustettin – Schneidemühl

Large Oval Postmark type III BArGe reference 7.3.2.0

SCHNEIDEMÜHL – STARGARD (POM.): Train number Z. 755; Trip date 17.1.37; Postmark period 1916-1945; Last day 25 January 1945

Field Post service mail addressed to Frankfurt/Main. Mail train returning via Küstrin. Departing Stargard (Pom.). Arrival in Schneidemühl station.

Details below from the Bahnpost Postmark catalogue

ROUTES	POSTMARK REFERENCE	POSTMARK PERIOD	NOTES
SCHNEIDEMÜHL – BELGARD	7.2.1.0	1923-1945	
SCHNEIDEMÜHL – CALLIES	7.1.2.0	1892	
SCHNEIDEMÜHL – CONITZ	5.2.3.2	1872	
SCHNEIDEMÜHL – FIRCHAU	7.3.6.0	1926-1943	First distribution date 20.5.1926
SCHNEIDEMÜHL – FLATOW	5.2.3.2	1871-1872	
SCHNEIDEMÜHL – NEUSTETTIN	2.2.3.0	1885	
SCHNEIDEMÜHL – NEUSTETTIN	7.1.1.0	1889	
SCHNEIDEMÜHL – POSEN	7.1.1.0	1899	
SCHNEIDEMÜHL – CZARNIKAU – ROGASEN	7.3.1.0	1913-1916	
SCHNEIDEMÜHL – STARGARD	7.1.1.0	1896-1925	
SCHNEIDEMÜHL – STARGARD	7.2.1.0	1934	
SCHNEIDEMÜHL – STARGARD (POM)	7.3.2.0	1916-1945	Last day of use 21.1.45
SCHNEIDEMÜHL – STARGARD (POMM.)	7.2.1.0	1926-1944	

120 HISTORY OF THE EASTERN RAILWAYS CONSTRUCTION AND EXPANSION

LARGE Oval Postmark type 7.III BArGe reference 7.3.4.0

BERLIN – SCHNEIDEMÜHL; Train number ZUG 346; Trip date 18.8.43; Postmark period 1936-1943 Diverse

Large Letter contains a set of Third Reich propaganda issues on an outbound mail train from Berlin to Schneidemühl addressed to Prague. Deutsches Reich, Foreign Letter rate up to 20g. = Pf.25 over by 1Pf. affixed MiNr.850-53

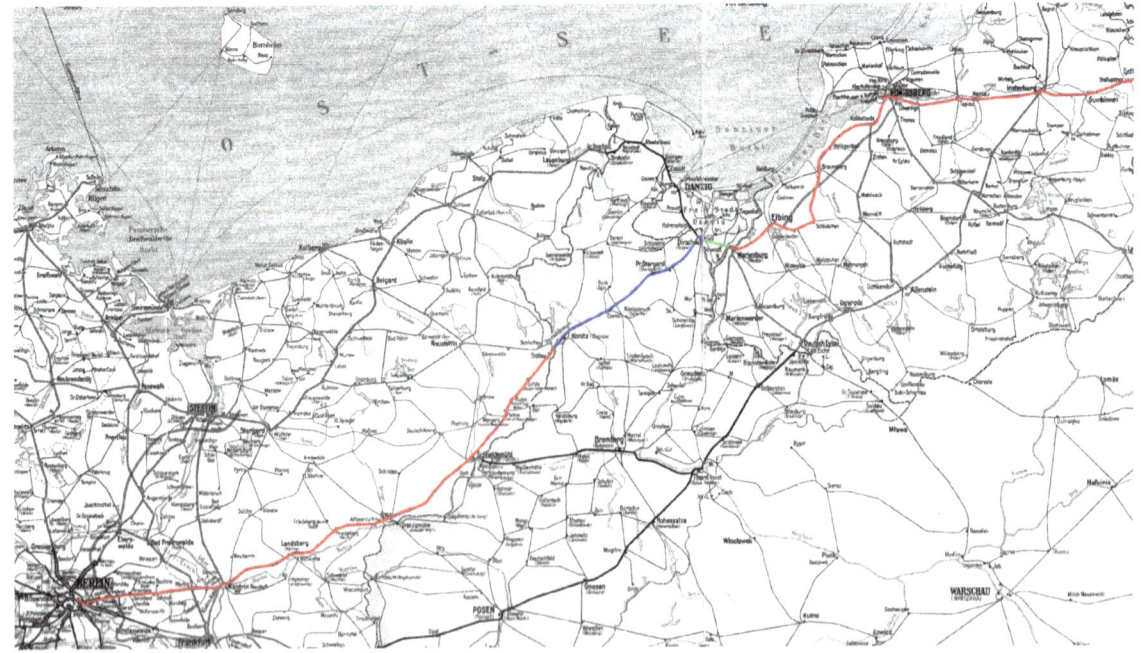

Railway line from Berlin to Schneidemühl

Timetable for route postmark Berlin – Schneidemühl (See page 120)

128 Berlin – Küstrin – Schneidemühl – Dirschau (-Königsberg [Pr]-Eydtkau)

Berlin – Coslin – Stolp – Danzig

38

These stations were connected between 1859 and 1870 by the Berlin Stettin Railway Company, which was one of the oldest railway companies in Germany. The extension from Stargard to western Pomerania took place at the instigation of the state, with an interest guarantee. This secured Prussia's right to take over if the routes were not profitable in the long term.

On 18 August 1856, the Berlin Stettin Railway Company was granted the concession for the Stargard-Belgard-Köslin and Belgard-Kolberg routes. These were opened on 1 June 1859. As a result of the extension of the line, the lease with the Stargard-Posen Railway Company on the Stettin-Stargard line ended on 1 January 1860.

Further connections were achieved with 67 kilometres from Cöslin to Stolp. The Stolp-Danzig route was completed by the Eastern Railway in 1852.

The route records a continous connection from Berlin Stettiner Railway Station to Stettin rail station - Stargard - Belgard - Cöslin - Stolp - Lauenburg (Pom) - Neustadt (Westpr) - Gotenhafen - Zoppot - Danzig.

Line Stamp Postmark type II BArGe reference 2.1.3.5

BERLIN - CÖSLIN; Tour/Trip 1T; Trip date 11 2; Script common Roman; Departing station / Data line / Arrival station, Postmark period 1860-1887

Pre-paid 1 Silbergroschen letter addressed to Belgard. Mail train outbound tour 1T from Berlin to Cöslin. Michel Ganzsachen U 17. Format B 148:115 mm embossed head 5.

Double Circle Railway Station Postmark

CÖSLIN BAHNHOF; Dispatch date 29 5 63 7-8V (before noon)

Pre-paid Prussian letter received for dispatch at the railway station post office addressed to Ueckermünde, west Pomerania. Michel Ganzsachen U 22.

Rail lines of Cöslin

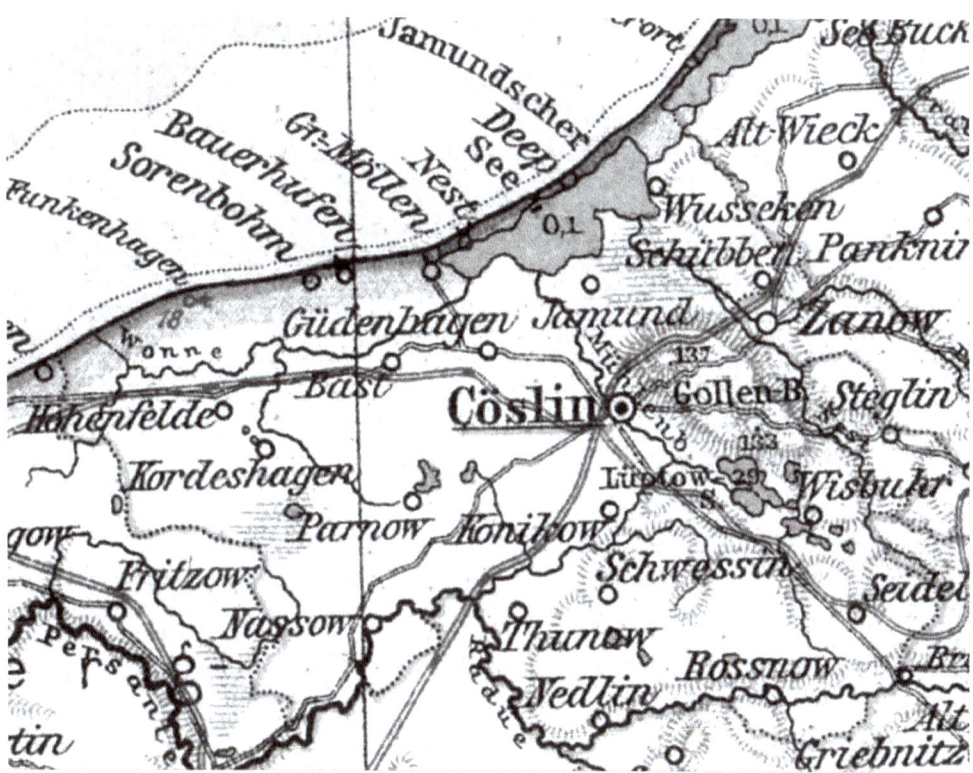

Line Stamp Postmark type 7.II BArGe reference 2.1.3.5

BERLIN – CÖSLIN; Tour/Trip IIT; Trip date 28 3; Script common Roman; Departing station / Data line / Arrival station; Postmark period 1860-1887

c.1863 addressed to Cöslin. Mail train outbound; Tour IIT from Berlin to Cöslin.

c.1900 Kaiserreich Head Post Office Köslin. Unused postcard.

Railway Postmarks of Cöslin and its railway history

The station was connected to the Berlin Danzig line in 1859. Neighbouring district towns were without a connection. In 1897, Prussian railway planners built an extensive network of narrow-gauge 750 mm railway lines connecting with Cöslin standard-gauge main station.

Line Stamp Postmark type II BArGe reference 2.1.3.5

CÖSLIN - BERLIN; Tour/Trip IIIR; Trip date 31 12; Script common Roman; Departing station / Data line / Arrival station; Postmark period 1860-1887

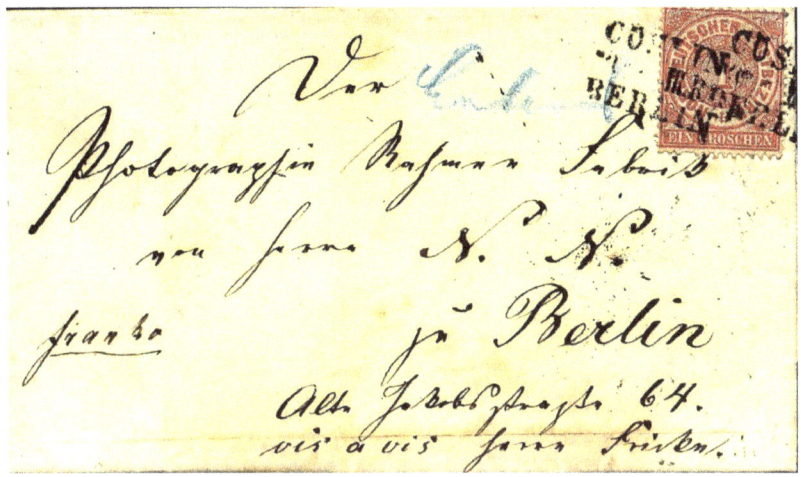

Post paid. For the North German Confederation, 1869-1870 was a period of political transition reflected on the postage stamp. A letter addressed to Berlin sent on mail train departing Cöslin - Belgard - Stettin rail station - arrival Berlin Stettiner Railway Station.

Railway Postmark catalogue for the route Cöslin:

ROUTES	REFERENCE	INDEX	POSTMARK PERIOD	NOTES
BERLIN – CÖSLIN	2.1.3.0		1867-1873	
BERLIN – CÖSLIN	2.1.3.5		1864-1878	
CÖSLIN – BERLIN	2.1.3.0		1872-1874	
CÖSLIN – BERLIN	2.1.3.5		1860-1887	
CÖSLIN – POLLNOW	7.1.1.0		1904-1906	
KÖSLIN – KOLBERG	7.1.2.0		1913-1927	
KÖSLIN – POLLNOW	7.1.1.0		1911-1922	
KÖSLIN – POLLNOW	7.1.1.1	a	1908	

Historical Development of Prussian-Era Steam Locomotives

THE DEVELOPMENT OF THE WET STEAM LOCOMOTIVE 1829-1859

The first German railway, connecting the cities of Nuremberg and Fürth, opened on 7 December 1835. A locomotive was purchased and shipped in 40 crates from England. The Adler, named after a German engineer, was built by Stephenson and Co. in Newcastle.

The Eagle

But the "Eagle" surely aroused far more amazement and admiration among the passengers of 1836 than the progress in locomotive construction did among the participants in the 100th anniversary of the railway. The leap in time from the slow stagecoach to the train "racing" at 40 km/h was enormous for the imaginations of the people of that time, and there were many citizens who were seriously worried about the health of the passengers at such speeds. With the use of the steam locomotive, a new era began in trade and transport. Countries and cities which previously could be reached only by a laborious, time-consuming journey moved closer together in a few years. Areas off the main roads were opened up to traffic, and the iron and steel industry received its first strong impetus.

For those who study the history of the locomotive in more detail, two facts are astonishing: all the basic features of the "Eagle" continued in locomotive design throughout the 20th century, and the meticulous engineering work of its humble beginnings was preserved without change in the powerful and reliable modern steam locomotives created by successive engineers.

In the first decade of the railway, locomotives similar to the Adler were used all over Germany. In the mid-19th century, the Prussian state began the construction of new state-owned lines in the direction of its eastern empire. This, in turn, gave rise to a large number of locomotive factories that emerged in quick succession, e.g. E. Keßler in Karlsruhe and Eßlingen, A. Borsig in Berlin, JA Maffei in Hirschau-Munich, G. Egestorff in Hanover, Henschel & Sohn in Kassel, L. Schwartzkopff in Berlin and R. Hartmann in Chemnitz.

The Crampton Built By Hanomag Linden Hannover

The Crampton locomotive (axle arrangement 2 A), named after its builder, was created as a competitive machine for high speeds. Its distinguishing feature was the drive axle with a large wheel diameter behind the firebox. While the Adler had a drive wheel diameter of only 1372 mm, the Crampton locomotive had a drive wheel diameter of 2134 mm, more than most of our new express train locomotives. As early as 1853, barely 20 years after the construction of the first railway in Germany, this locomotive reached a speed of

2 A-Zweizylinder Naßdampf-Schnellzuglokomotive Bauart Crampton, der Hannoverschen Staatsbahn gebaut 1857 von HANOMAG Hannover-Linden. 1857.

The Crampton Locomotive on Hannover and Eastern Railways

Specifications: heel arrangement 4-2-0 or older variant 6-2-0. Bogie wheels 1.07 m diameter, driving wheels 1.83 m diameter. Cylinders 381 mm × 559 mm. Weight 26 long tons 5 cwt.

120 km/h without a load and 80 km/h with a trailer load of 50 t. The locomotive (boiler heating surface 68.6 m²) was built by Maffei-Munich in 1853 for the Pfalzbahn and by Hanomag-Linden Hannover in 1859.

Notable features were a low boiler and large driving wheels. The crux of the Crampton patent was that the single driving axle was placed behind the firebox so that the driving wheels could be very large. The Crampton locomotive was characterised by the semi-cylinder vaulted standing boiler ceiling, which can still be found on most locomotives today.

Since the Crampton locomotive also carried the steam dome on the stand-up boiler ceiling, a very large part of the locomotive's weight rested on the drive axle, which was able to transfer large traction forces accordingly. From 1855 onwards, the Crampton locomotive was used in express train service because of its smooth running, but from 1860 onwards, the frictional weight of the one driving axle was generally no longer sufficient, so a switch was made to the multi-coupled locomotive for express and passenger train traffic.

Prussian Locomotives on the Mountain routes

At a relatively high stage of development, about 10 years after the construction of the first railways in Germany, the planners and engineers built the first triple-coupled locomotives, initially again mainly for mountain routes and freight train traffic.

The locomotive with an outer frame

In the years 1855 to 1860, the almost exclusively used locomotive fuel, coke, was replaced by hard coal. At the same time, attempts were made to increase the boiler output by increasing the grate area (the key combustion mechanism) by placing the grate at an angle and lengthening it beyond the rear axle. In 1870, the tilting grate and the boiler scale separation in the dining dome were already known in their first designs.

An example is the three-axle locomotive below, built by Schwartzkopff-Berlin in 1867. The frame of this machine was placed on the outside, a design that had the advantage that the firebox could be made wider than with an internal frame and was widespread in Austria and Hungary at that time. The disadvantage of the outer frame was the difficult accessibility of the inner engine. The locomotive remained the most common and most economical freight locomotive in the second half of the 19th century. At that time, the need for a four-way coupled machine existed only for steep mountain railways. In addition, the greater length and weight of this type of construction initially made it difficult to negotiate bends.

After 1890, longer and heavier freight trains generally required four axles to be coupled. The construction of these machines took place at a time when the introduction of the composite process for the steam engine brought considerable progress in terms of heat economy and performance to locomotive construction.

The Prussian D-ZUG Locomotives of the Eastern Railway

40

The endeavour to improve economic efficiency, together with higher boiler pressure, led at the end of the 1870s to no longer releasing the steam in one cylinder but instead using its working capacity successively in a high-pressure and a low-pressure cylinder. This so-called composite effect was based on attempts by Professor Bauschinger from Munich. In 1865 he discovered that the greatest heat losses in the steam cylinder are caused by inlet condensation, i.e. by the cooling and partial precipitation of the saturated steam on the colder cylinder walls. At high pressure, the greater the temperature gradient in the steam engine, the greater the losses. The results of these tests then gave the Swiss engineer Mallet the idea.

The development of the compound locomotive at the Prussian State Railways is inextricably linked with the name of August von Borries. The locomotives he built, in which the composite effect was always maintained by an automatic start-up valve except when starting, created fuel savings of 15 - 16% compared to twin locomotives of the same size. If the compound principle was used in 2-cylinder locomotives, the low-pressure cylinder was built larger than the high-pressure cylinder so that the same power was available on both sides of the locomotive. This design led to unequal engine masses and

The 2 B1 n4v express locomotive

uneven running. For high-speed locomotives, 4-cylinder compound locomotives were therefore developed, first in France and then in Germany.

In the second half of the 19th century, the express and passenger train service was served almost exclusively by double-coupled locomotives, either in the axle arrangement 1B, which was often used for passenger locomotives, or type 2B, used for express locomotives with two cylinders and single or double steam expansion. The steadily increasing demands on locomotive performance and running safety led to a steady increase in the number of express locomotives. How fierce the competition was between the individual types can be seen from the fact that between 1907 and 1910, at a time when the superiority of superheated steam operation was recognised, there was still a 2'B1' n4v locomotive being built; it had a boiler pressure of 14 kg/cm 2, the driving wheel diameter was 1980 mm, and the coupling axle pressure had increased to 16.5 t. The 4-cylinder locomotives had the advantage of keeping the operation free of disruptive vibrations by rotating the two right and left engines in opposite directions. Even with the similarly built predecessor of the S 9, the S 7 locomotive, speeds of up to 143 km/h had been achieved. In terms of running technology, these locomotives were already able to cope with the demands of high-speed traffic.

The 2-B-1 Express Locomotive – Prussian State Railway P 4

41

The Prussian P 4 was a derivative of the P 4.1 (Hanover variant) and the second superheated steam locomotive in the world. The engine was based on that of the Class P 4.1 that Hanomag had produced in large numbers since 1892. It had slightly larger wheels and, due to its new design, significantly fewer heating tubes. The superheater and the steam engine were entirely independent designs.

Manufacturer Henschel; Year of manufacture 1898; Number of locomotives 1; Retired 1921; Wheel arrangement 4-4-0; Axle arrangement 2'B h2; Type P 24.45; Track gauge 1,435; Length over buffers 16,411 mm; Overall wheelbase 7,400 mm; Empty weight 44.6 t; Service weight 49.1 t; Adhesive weight 31.0 t; Axle load 15.5 t; Top speed 90 km/h; Indicated Power 675 kW (900 PS); Coupled wheel diameter 1,750 mm; Carrying wheel diameter 1,000 mm; No. of cylinders 2; Cylinder bore 460 mm; Piston stroke 600 mm; Boiler overpressure 12 bar; No. of heating tubes 141; No. of smoke tubes 1 fire tube; Grate area 2.32 m2; Radiative heating area 8.9 m2; Tube heating area 76.2 m2; Superheater area 21.0 m2; Evaporative heating area 85.1 m2; Tender pr 3 T 15; Water capacity 15.0 m3.

In 1898, a one-off was delivered by Hanomag to the Prussian state railways. The economy of the superheated system was soon proven the following year during trial runs from Kassel. Apart from a short stay at Halle, the engine was assigned to Kassel as Cassel 131 and, from 1906, as P 4 Cassel 1846. In 1921, after the First World War, the engine was mothballed, along with many other machines of similar class. With its sectioned boiler, the P 4 stood for a long time in the Transport and Construction Museum, part of the Museum of the Present at Hamburger railway station in Berlin.

PRUSSIAN STATE RAILWAYS – LOCOMOTIVE SERIES TYPE P 8

The Prussian P 8 was a passenger locomotive operated by the Prussian state railways. It was a steam locomotive with a 4-6-0 wheel arrangement built from 1906 to 1923 by the Berliner (previously Schwartzkopff) and twelve other German factories. It was designed by Robert Garbe and intended as a successor to the Prussian P 6, which was regarded as unsatisfactory. It was an advocate of the simplest possible designs, a straightforward, superheated steam engine, with a two-cylinder driving gear.

Manufacturer Hohenzollern, Schwartzkopff; Years of manufacture 1902–1910; Retired about 1950; Wheel arrangement 2-6-0; Axle arrangement 1'C h2; Track gauge 1,435 mm; Length over buffers 17,608 mm; Service weight 57.1 t (56.2 long tons; 62.9 short tons); Adhesive weight 44.6 t (43.9 long tons; 49.2 short tons); Axle load 15.2 t (15.0 long tons; 16.8 short tons); Top speed 90 km/h; Indicated power 755 kW (1,012 hp); Driving wheel diameter 1,600 mm; Leading wheel diameter 1,000 mm; No. of cylinders 2; Cylinder bore 540 mm; Piston stroke 630 mm; Boiler overpressure 12 bar; Grate area 2.28 m2; Superheater area 41 m2; Tender pr 2'2' T 16; Water capacity 16.0 m3 (3,500 imp gal; 4,200 UK gal)

The locomotives were authorised to travel at up to 90 km/h (56 mph), a speed which could not be attained in practice due to its poor operational qualities. The smokebox superheater installed on the first machines was soon replaced by a smoke tube superheater. In all, 275 engines of this class were built up to 1910. 110 examples had to be handed over after the First World War as reparations. 163 locomotives were taken over by the Deutsche Reichsbahn as DRG Class 37.0-1, where they were allocated the running numbers 37 001-163. A few of these have survived and are now in railway museums in Germany and Poland.

The Prussian P 6s were retired by about 1950. The few engines left after the Second World War were no longer employed by the Deutsche Bundesbahn or the Reichsbahn. The locomotives taken over by the Polish State Railway (PKP) were given the designation Oi1.

Prussian State Railways Locomotive Class S 2

42

Express train locomotives of the Prussian state railway were classified as S 2. These locomotives were the first of the Prussian state railway with an axle formula 2'B. In contrast to most of the other S 2s, they were built in a composite construction. They were designed by August von Borries. A special feature was the so-called Hanover bogie. The load was transferred directly to the longitudinal suspension spring via the two sliding blocks of the spring-loaded return device without loading the bogie. This principle was retained until the end of the steam locomotive era.

2/5 gek. Schnellzuglokomotive mit 4 Zylindern mit Pielocküberhitzer.

Wheel arrangement 4-4-0; Axle arrangement 2'Bn2v; Type: 2'Bn2v; Gauge 1435 mm; Length over buffers 15,380 mm; Service mass 45.0t; Friction mass 28.0 t; Wheelset mass 17.6 t; Top speed 100 km/h; Driving wheel diameter 1,960 mm; Impeller diameter front 980 mm; Number of cylinders 2; Cylinder diameter 450/650 mm; Piston stroke 600 mm; Boiler overpressure 12 bar; Grate area 2.07m²; Tender pr 3 T 10.

Only two of the Hanover types that Henschel had built in 1890 were put into service. The vehicles were equipped with type 3 T 10.5 tender. As they gave poor performance, like the other older composite machines, this type was no longer built. The locomotives were retired in 1912.

From 1906, a total of 26 S 2s were converted into a composite design and then classified as S 3. A locomotive delivered with a Lentz corrugated tube boiler failed to achieve the desired performance. It pulled a 210-ton train up a gradient of 5 per thousand at a speed of only 50 km/h. In 1923 the Deutsche Reichsbahn took over seven S 2s converted to composite design as 13 001 - 007 in their drawing plans for steam locomotives. Only one locomotive was included in the 1925 redesignation plan as 13 001, and it was taken out of service by 1926. Some of the redesigned locomotives were handed over to Poland in the 1920s.

PRUSSIAN STATE RAILWAY 4-4-4 EXPRESS LOCOMOTIVE TYPE S 8 (PIELOCK OVERHEATER)

The Prussian S 9 was an express steam locomotive on the Prussian state railways, first built in 1908. It had a 4-4-4-2 (Atlantic) wheel arrangement and a four-cylinder

2/5 gek. Schnellzuglokomotive mit 4 Zylindern mit Pielocküberhitzer.

Numbers DRG 14 001, 002, 031; Quantity 99 (2 were S8s); Manufacturer Hanomag; Years of manufacture 1908; Retired 1926; Axle arrangement 4-4-2 for S 8: 2'B1' h4v (S 9: 2'B1' n4v); Type S 8: S 2/5 h4v for (S 9: S 2/5 n4v); Track gauge 1,435 mm; Length over buffers 21,860 mm; Length 13,110 mm (excl. tender); Overall wheelbase 10,750 mm; Empty weight 68.0 t; Service weight 74.7 t; Adhesive weight 33.0 t; Axle load 16.5 t; Top speed 110 km/h; Coupled wheel diameter 1,980 mm; Leading wheel diameter 1,000 mm; Trailing wheel diameter 1,250 mm; No. of cylinders 4; Piston stroke 600 mm; Boiler overpressure 14 bar; No. of heating tubes 272; Heating tube length 5,200 mm; Grate area 4.00 m2; Radiative heating area 14.10 m2; Tube heating area 222.00 m2; Evaporative heating area S 8: 229.71 m2 for (S 9: 182.54 m2); Tender pr 2'2' T 21.5/30/31.5; Water capacity 21.5/30.0/31.5 m3

compound engine. It was developed by the firm Hanomag in Hannover, which delivered a total of 99 engines of this class. Two locomotives (Hannover 903 and 905) were fitted with superheated boilers in 1913 and 1914, respectively, and reclassified as S 8s.

Although, at the time, superheated technology was widespread, the state of Prussia still wanted the saturated steam engines delivered by Hanomag. The locomotives procured as a result had a very powerful boiler and, at 4 m2, the largest grate area of any Prussian steam locomotive. The quantity of steam generated was, however, more than the high-pressure cylinders could cope with. As a result, the performance of the S 9 was little better than the considerably smaller superheated locomotive, the Prussian S 6.

After 1919, 17 locomotives had to be handed over to Belgium and four to France. Only three of them, the two S 8s and a saturated steam engine, were taken over by the Deutsche Reichsbahn as DRG Class 14.0. The two S 8s were given the numbers 14 001 and 14 002; the S 9 (Essen 907) number 14 031. All three were retired by 1926.

Prussian State Railways – Locomotive Series T 8

43

The Prussian T 8 was a six-coupled superheated goods tank locomotive of the Prussian state railways. It was originally intended for suburban passenger service in Berlin and use on branch lines. Due to their poor running qualities, they were demoted to shunting and short-distance goods train service.

Between 1906 and 1909, 100 locomotives were built, of which 80 were still in service with Deutsche Reichsbahn in 1923, and 78 in 1925, when they were renumbered in class 89.0 as 89 001 to 078; due to their poor performance, they were soon sold to private railways.

After World War I, ten locomotives were ceded to Poland, where they became PKP class TKh3.

In 1938, two locomotives were taken back into stock when private companies operating on secondary routes were nationalised; rather than restoring the locomotives' old numbers, they were allocated new ones: 89 1001 and 1002.

Manufacturer MBG Breslau, O&K and Hanomag; Year) of manufacture 1906–1909; Quantity 100; Retired 1965; Axle arrangement C h2t; Track gauge 1,435 mm (4 ft 8 1/2 in); Length over buffers 9,460 mm (31 ft 1/2 in); Service weight 45.6 tonnes (44.9 long tons; 50.3 short tons); Adhesive weight 45.6 t; Axle load 15.5 tonnes (15.3 long tons; 17.1 short tons); Top speed 60 km/h (37 mph); Indicated Power 210 kW (286 PS; 282 hp); Driving wheel diameter 1,350 mm (4 ft 5 1/8 in); Cylinder bore 500 mm (19 11/16 in); Piston stroke 600 mm (23 5/8 in); Boiler overpressure 12 bar (1.20 MPa; 174 psi); Grate area 1.51 square metres (16.3 sq ft); Superheater area 17.90 square metres; Evaporative heating area 68.50 m2; Brakes Hand brake, later compressed air brake.

The last DR locomotive was retired in 1965; the last Deutsche Bundesbahn locomotive (a works locomotive) was retired in 1964.

PRUSSIAN STATE RAILWAYS – LOCOMOTIVE SERIES TYPE T 16

PRUSSIAN STATE RAILWAY
CLASS T16 0-10-0 HEAVY SHUNTING LOCOMOTIVE No. 094 564-5 BUILT 1914 BY F. SCHICHAU GmbH. ELBING

Axle arrangement E h2t; Length over buffers 12,500 mm (41 ft 1/4 in); Axle load 16.5 tonnes (16.2 long tons; 18.2 short tons); Type Gt 55.17; Service weight 75.6 tonnes (74.4 long tons; 83.3 short tons); Adhesive weight 75.6 tonnes (74.4 long tons; 83.3 short tons); Top speed 40 km/h (25 mph)

The Prussian T16 locomotives were ten-coupled superheated freight tank locomotives of the Prussian state railways. They were later renumbered in the 94.2-4 class by the Deutsche Reichsbahn.

The development of the T16 locomotives was influenced by the ideas of Karl Gölsdorf. The design included three Gölsdorf axles - the first, third, and fifth - which enabled the elimination of the articulated frame that had been used on the Prussian T 15. While mainly bought for banking (pushing trains up steep gradients), they were also used for freight trains and shunting. They proved to be more powerful and economical than the T 15 class.

The Deutsche Reichsbahn renumbered the T 16s 94 201-467, although the last three were actually T 16.1s; this error was corrected in 1934. In addition, one T 16 locomotive had been renumbered as a T 16.1 in error, but 94 501 was withdrawn in 1931 before the error was corrected. Locomotives 94 462-464 came from Alsace-Lorraine. Eight T 16s remained in Alsace-Lorraine and eventually became SNCF 1-050.TA.101 to 112. Many of the DR locomotives were retired in the 1930s.

During World War II, a number of Polish locomotives and one Belgium locomotive were taken into stock as 94 468-490. One Belgian locomotive was in the Soviet zone at the end of the war and became 94 1811; it formed part of the newly formed East German Deutsche Reichsbahn fleet.

The Deutsche Reich Railway (DRG) Series 43

44

The German locomotives of DRG Class 43 were standard goods train engines with the Deutsche Reichsbahn. This was the second class that was built on the standard steam locomotive principle. According to the first classification scheme of the DR's Standardisation Bureau, 2-10-0 goods train locomotives were to be procured with a 20-ton axle load. To achieve this, a two-cylinder class (Class 43) and a three-cylinder class (Class 4) were envisaged because it was not yet known which configuration would be more economical. Ten examples of each class were procured in parallel. The Class 43 locomotives were supplied by Henschel and Schwartzkopff.

In trials, it was discovered that the Class 43 was more economical when operating at powers under 1500 PSi. As a result, a further 25 of Class 43 were ordered up to 1928. However, due to the increase in goods train speeds in the 1930s, the Class 44 was given priority thereafter because, in addition to its economy when running at higher speeds

and the better running qualities of its three-cylinder driving gear, it was also cleared for running at up to 80 km/h. As a result, no more Class 43s were ordered.

The Class 43s, which had operating numbers 43 001 - 43 035, all remained with the DR in East Germany after the Second World War. In 1960, the remaining engines were once again modernised; the powerful boiler even enabled goods trains greater than the maximum allowable load to be hauled. This led, however, to frame damage that the Reichsbahn could not repair, so the machines were rapidly retired, the last one being taken out of service in 1968 at Cottbus locomotive depot.

The Deutsche Reich Railway Company Class 24 (Sketch) 45

The Deutsche Reich Railway Company DRG Class 24 steam engines were German standard steam locomotives built for the Deutsche Reichsbahn between 1928 and 1939 to haul passenger trains. These engines, nicknamed the "Prairie Horse" (Steppenpferd) were developed especially for the long, flat routes of the Eastern Railway in West and East Prussia. These locomotives were operational with a boiler overpressure of 245.1 N/cm2 (355.5 psi) but were rebuilt in 1952.

The locomotives were built by the firms Schichu, Linke-Hofmann and Borsig; a total of 95 were delivered for service.

The Deutsche Bundesbahn took over 38 locomotives and retired them by 1966. Thirty-four locomotives remained in Poland after the Second World War, where PKP classified them as Oi2. They served until the last one was withdrawn in 1976. One of the preserved locomotives in Germany, no. 24 083, had been in service in Poland.

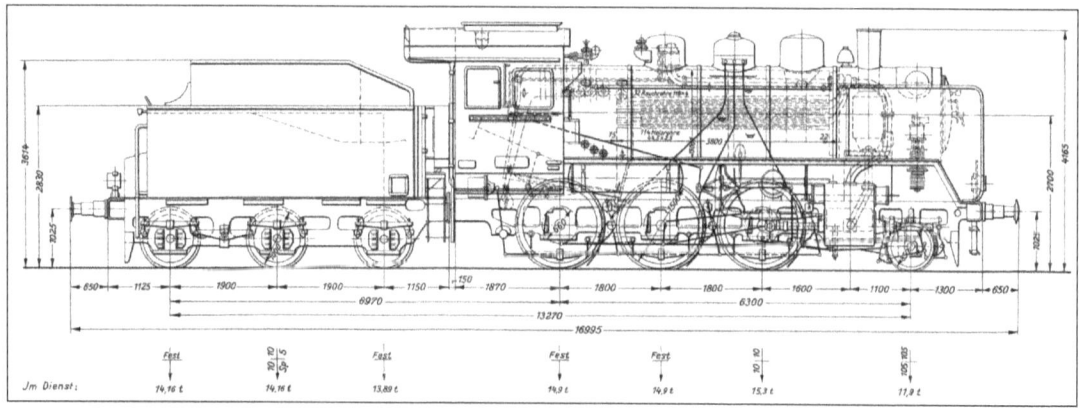

1'C h2 passenger locomotive, class 24, tender 3 T 16 (version with Krauss-Helmohltz steering frame)

Design type axel 1'C h2

Construction year 1926

Machines built in 95

Last year of service 1966

Driving and coupling wheel 1500 mm

Impeller Ø front - 850 mm

Impeller Ø rear -

Fixed wheelbase 1800 mm

Total wheelbase 6300 mm

Length over buffers 16,955 mm

Top speed 90 km/h

Power 920 Psi

Boiler overpressure 14 kgf/cm^2 (Kilograms per sq. centimetre)

Grate surface 2.04 m^2

Firebox heating surface 8.7 m^2

Evaporation heating surface 104.48 m^2

Overheating heating surface 37.340 m^2

Cylinder diameter 500 mm

Piston stroke 660mm

Axle Load maximum 15.1 Mp

Locomotive friction load 45.2 Mp

Locomotive friction load 45.2 Mp

Tender 3 T 16 and 3 T 17

THE DEUTSCHE REICH RAILWAY COMPANY EXPRESS CLASS BR 01-10 (SKETCH)

The Deutsche Reich Railway Company's BR 01 steam locomotives were the first standardised steam express passenger locomotives built by the unified German railway system. They had the 4-6-2 Pacific wheel arrangement in the Whyte notation or 2'C1' h2 in the UIC classification. The idea of standardisation was that it would reduce maintenance costs. The firms AEG and Borsig, who were the main manufacturers of these engines, together with Henschel, Hohenzollern, Krupp and Schwartzkopff, delivered a total of 231 of these engines between 1926 and 1938.

Originally, the Class 01's top speed was restricted to 120 km/h. In order to raise this to 130 km/h, the diameter of the leading wheels was changed from its original 850 mm to 1,000 mm on locomotives from operating number BR 01 102.

The development of high-speed engines continued in 1935 with the appearance of the Class BR 05 locomotive, which was also designed to standard locomotive principles, although only three examples were produced. The Class 05 was permitted to travel regularly at a top speed of 175 km/h and set the world speed record for steam engines of 200 km/h. This record was later officially beaten by the English "Mallard", an LNER Class A4 locomotive.

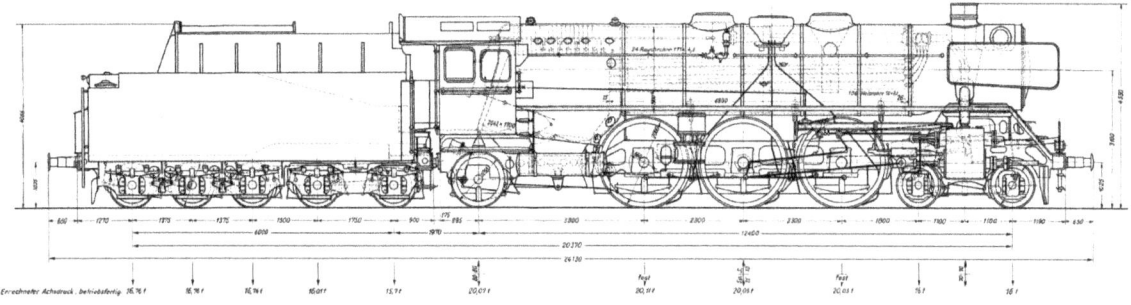

2'C 1 'h3 express locomotive, class 01¹⁰, tender 2'3 T 38 (formerly St)

Design type axel 2'C 1 'h3

Construction year 1939

Machines built in 55

Driving and coupling wheel 2000 mm

Impeller Ø front - 1000 mm

Impeller Ø rear - 1250 mm

Fixed wheelbase 4600 mm

Total wheelbase 10400 mm

Length over buffers 24 130 mm

Top speed 140 km/h

Power 2350/2470 *Psi

*with oil firing

Boiler overpressure 16 kgf/cm²
(Kilograms per sq. centimetre)

Grate surface 3.96 m²

Firebox heating surface 16.9 m²

Evaporation heating surface 206.51 m²

Overheating heating surface 96.15 m²

Cylinder diameter 500 mm

Piston stroke 660mm

Axle Load maximum 20.2 Mp

Locomotive friction load 60.4 Mp

Locomotive service load 110.8 Mp

Tender 2'3 T 38

THE DEUTSCHE REICH RAILWAY (DRG) GOODS CLASS 80 (SKETCH)

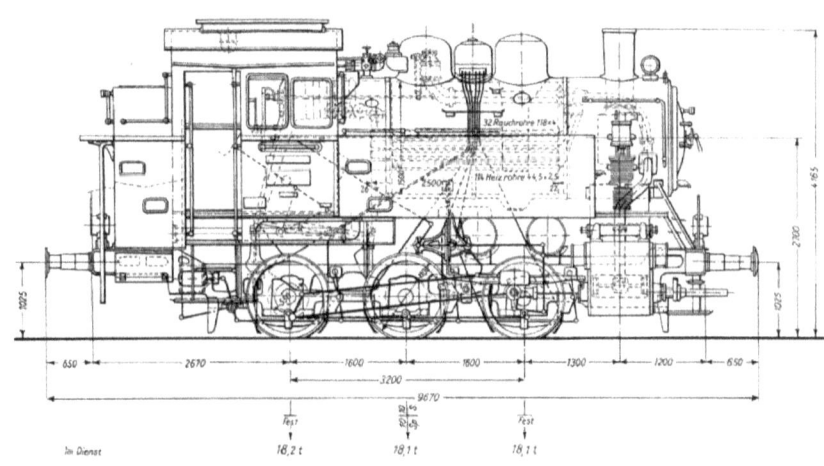

C h 2 freight train tank locomotive, class 80

Design type axel C h2

Construction year 1927

Machines built in 39

Driving and coupling wheel Ø 1100mm

Type Gt 33.17

Height 4,165 mm

Wheel arrangement 0-6-0T

Total wheelbase 3200 mm

Length over buffers 9670 mm

Top speed - 45 km/h

Power 575Psi

*with oil firing

Boiler overpressure 17 kgf/cm² (Kilograms per sq. centimetre)

Grate surface 1.54 m²

Firebox heating surface 6.6 m²

Evaporation heating surface 69.62 m²

Overheating heating surface 25.50 m²

Cylinder diameter 450 mm

Piston stroke 550mm

Axle Load maximum 18.2 Mp

Locomotive friction load 54.4 Mp

Locomotive service load 54.4 Mp

Track gauge 1435 mm

Retired 1977

Brakes - Direct-release Knorr compressed-air brakes

Parking Brakes K-GP mZ counterweight handbrake

THE DEUTSCHE REICH RAILWAY (DRG) GOODS CLASS 80

The Class 80 tank engines were German standard locomotives with the Deutsche Reichsbahn. They were intended to replace the ageing, rickety state railway line engines performing shunting duties in their dotage at large stations.

Between 1927 and 1928, a total of 39 vehicles were produced, having been built in the locomotive factories of Jung in Jungenthal and Union Gießerei in Königsberg, Wolf and Hohenzollern. With the development of the Class 80, a relatively economical and simple locomotive class, it was hoped that the cost of shunting would come down.

Prior to the Second World War, they had been on duty primarily in the area of Leipzig. (Their work included the shunting of post vans.) After the war, they worked in Cologne. Following the division of Germany, 22 units went into the DR in East Germany. A total of

17 locomotives with the Deutsche Bundesbahn continued in service on various routes for a short period, and some continued in service until 1968.

In the Federal Republic of Germany, the last Bundesbahn engine was taken out of service in 1965. Several examples survived in the Ruhrgebiet until 1977 as industrial locomotives with the Ruhrkohle Company. A total of seven locomotives of this class have been preserved.

THE DEUTSCHE REICH (DRG) GOODS CLASS 85

The Class 85 was a German goods train tank locomotive engine and standard locomotive with the Deutsche Reichsbahn. The heavy three-cylinder tender machines of the BR 85 (type 1'E1' h3-t) were designed for this purpose. They were a replacement for the BR 95 (pr. T 20) which was no longer an option despite the adequate performance of this model. The construction of the BR 85 benefited from various assemblies from the standard program: the chassis, engine and control came from the BR 44 and the boiler from the BR 62.

Goods train tank locomotive series 80: 81: 84: 86: 87: 89.0

In 1931, the DRG ordered ten locomotives from the firm of Henschel that were taken into the fleet as numbers 85 001 to 85 010. Class 85 was intended for hauling passenger and goods trains, as well as mail wagons on night runs. They were, however, also employed as pusher locomotives on the Höllentalbahn in the Black forest. Thanks to this engine, the Höllental Railway could do away with rack railway operations from 1933.

The running gear and the superheated system were taken from Class 44. The boiler with a few minor alterations was the same as that of Class 62. All the locomotives were stabled at the Freiburg shed. One of the DRG engines, number 85 004, was recorded as lost in the Second World War. This missing locomotive would have fallen into the hands of the Russians when it was possibly supplying Wehrmacht troops in the east.

All the engines were in operation in the Black Forest until 1961, the year the route was converted from experimental electrical operations with 20 kV/50 Hz lines to the usual Deutsche Bundesbahn standard of 15 kV 16.7 Hz AC. One engine, number 85 007, was still in service in Wuppertal until the end of the year, but they were all retired by the beginning of the next year.

THE DEUTSCHE REICH (DRG) GOODS CLASS 85 (SKETCH)

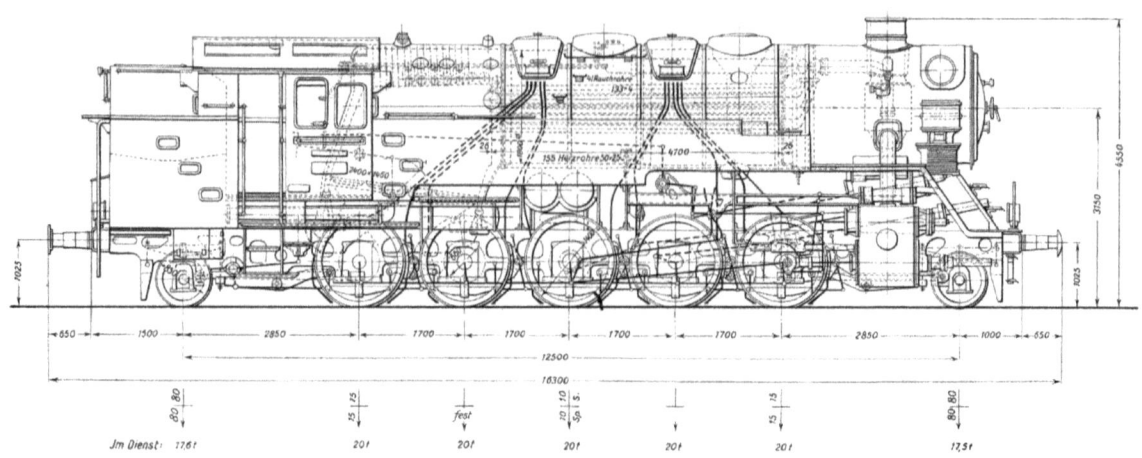

1'E 1' h 3 freight train tank locomotive, class 85

Design type axel 1'E 1 'h3

Construction year 1932

Machines built in 10

Manufacturer Henschel & Sohn

Driving and coupling wheel Ø 1400mm

Height 4,550 mm

Wheel arrangement 2-10-2T

Total wheelbase 12500 mm

Length over buffers 16,300 mm

Top speed - 80 km/h

Power 1500Psi

Fuel - Coal 14.5 tonnes

Water capacity 14 m3 (490 cu ft)

Retired 1961

Boiler overpressure 14 kgf/cm² (Kilograms per sq. centimetre)

Grate surface 3.55 m²

Firebox heating surface 15.0 m²

Evaporation heating surface 195.95 m²

Overheating heating surface 72.50 m²

Cylinder diameter 600 mm

Piston stroke 660mm

Axle Load maximum 20.1 Mp

Locomotive friction load 99.7 Mp

Locomotive service load 133.6 Mp

Track gauge 1435 mm

Locomotive width 3,050 mm

Impeller Ø front 850 mm

Impeller Ø rear 3400 mm

THE DEUTSCHE REICH (DRG) GOODS CLASS 86 (SKETCH)

The DRG Class 86 was a standard goods train tank locomotive with the Deutsche Reichsbahn Company. It was intended for duties on branch lines and was delivered by almost all the locomotive building firms working for the Reichsbahn. From 1942, it was built in a simplified version as a "transitional war locomotive". The most obvious changes were the omission of the second side windows in the cab and the solid disc carrying wheels.

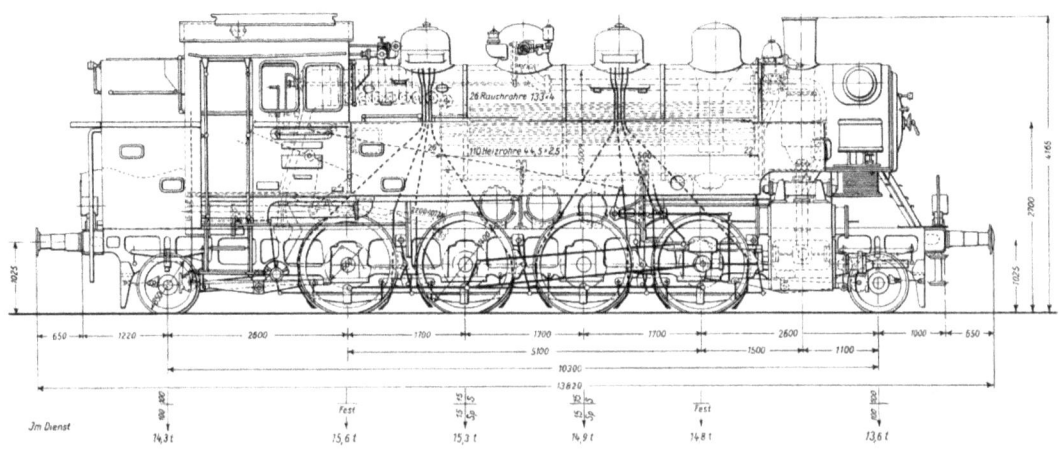

Axle arrangement 1'D1' h2t; Wheel arrangement: 2-8-2T; Leading wheel diameter: 850 mm; Driving wheel diameter: 1,400 mm; Trailing wheel diameter: 850 mm; Axle load 15.2 tonnes (15.0 long tons; 16.8 short tons); Service weight 88.5 tonnes Top speed 70–80 km/h (43–50 mph); No. of cylinders 2; Piston stroke 660 mm (26 in); Grate area 2.39 m2 (25.7 sq ft); No. of smoke tubes 26; Valve gear outside Walschaerts valve gear; Fuel Coal: 4.0 tonnes (3.9 long tons; 4.4 short tons)

Almost all German locomotive factories took part in building these engines, 775 examples being produced in the period from 1928 to 1943. They operated predominantly on the routes in Germany's Central Uplands, so the first 10 units were given Riggenbach counter-pressure brakes. Twenty locomotives were destroyed during the Second World War; lightly damaged engines were repaired. Of the original 775 units, 175 went to the GDR railways and 385 to the Deutsche Bundesbahn.

In the GDR railways, the 86s were mainly stationed at the Aue engine shed (over 50 engines) for the surrounding Ore mountain routes. Some DR engines stationed at the Heringsdorf shed on the island of Usedom were even given smoke deflectors. In their latter years, they were often used just as heating engines. Their last duties were on the stub line from Schlettau to Crottendorf, where they ended steam services in 1988.

www.ingramcontent.com/pod-product-compliance
Lightning Source LLC
Chambersburg PA
CBHW042019090526

44590CB00029B/4330